THE COSMIC CODE

QUANTUM PHYSICS AS THE LANGUAGE OF NATURE

Heinz R. Pagels

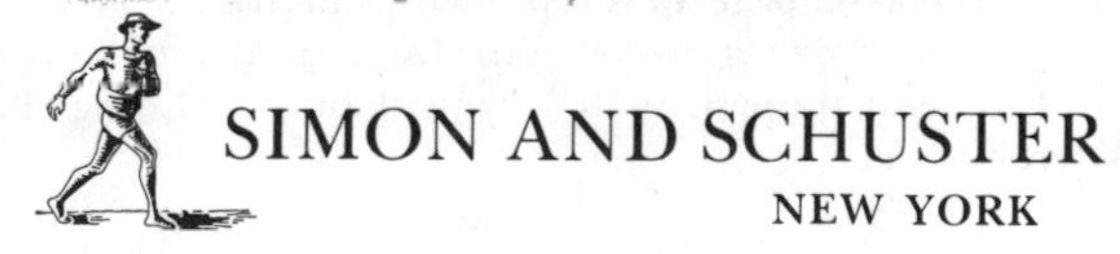

SIMON AND SCHUSTER
NEW YORK

Published by Simon and Schuster
A Division of Gulf & Western Corporation
Simon & Schuster Building
Rockefeller Center
1230 Avenue of the Americas
New York, New York 10020

SIMON AND SCHUSTER and colophon are trademarks of Simon & Schuster
Designed by Irving Perkins Associates
Illustrations by Matthew Zimet
Manufactured in the United States of America

1 2 3 4 5 6 7 8 9 10

Library of Congress Cataloging in Publication Data

Pagels, Heinz R., date
The cosmic code.

Bibliography: p.
Includes index.
1. Quantum theory. 2. Particles (Nuclear physics)
3. Science—Philosophy. I. Title.
QC174.13.P33 530.1'2 81-16525
ISBN 0-671-24802-2 AACR2

acknowledgments

In November 1977 I attended a symposium at Columbia University to honor Professor Isidor Isaac Rabi, an experimental physicist of the Los Alamos generation, a Nobel laureate, and a statesman of science. He was a founder of Brookhaven National Laboratory and the European Organization for Nuclear Research (CERN). After a day of speeches by his colleagues, Rabi made a few remarks of his own. He chided physicists for failing to bring the excitement of physics to a larger public and said that they had done less than science fiction writers to communicate the spirit of science. Upon hearing his remarks I resolved to write this book —a resolution supported by my friend John Brockman's instinct for intellectual excitement. With John's gentle prodding I wrote a book proposal.

Many people, all friends, have made suggestions regarding style and content which have found their way into the final product—Kathryn Burkhart, Ashton Carter, Sidney Coleman, Rodney Cool, Gerald Feinberg, Daniel Greenberger, Mark Kac, Tony King, Linda Hess, Emily McCully, Richard Ogust, Hilary Putnam, and especially Eugene Schwartz and Arthur Miller whose criticisms helped me enormously. Many of my views on the problem of quantum reality grew out of delightfully humorous and informative discussions with Nicholas Herbert. I was also fortunate that the writing of this book coincided with the centennial year of Einstein's birth, 1979. In that year I attended three centennial symposia, one held at the Institute for Advanced Study in Princeton, N.J., another in Jerusalem sponsored by the Israeli Academy of Science and the van Leer Foundation, and a third in

New York City sponsored by the New York Academy of Sciences. I benefited greatly from lectures at these meetings, especially those by Daniel Bell, Jeremy Bernstein, Erik Erikson, Loren Graham, Gerald Holton, Martin Klein, Arthur Miller, Abraham Pais, Wolfgang Panofsky, Dennis Sciama, Irwin Shapiro, Steven Weinberg, and John Wheeler. Articles by A. Pais and G. Holton on the early work of Einstein were especially useful. The second part of the book has been influenced by U. Amaldi's CERN article on accelerators and scientific culture as well as the 1977 *Dedalus* article of Steven Weinberg. The third part of the book grew out of conversations with my friend Joseph H. Hazen.

It was my good fortune to find Matthew Zimet, who illustrated the book. His novel and humorous drawings do much to ease the burden of the text by delighting mind and eye.

A scientist's greatest asset, the ability to concentrate productively on specific problems until they yield, can become a liability when he attempts to communicate his ideas to a nonscientist. Therefore my deepest debt is to my editors, Alice Mayhew and Catherine Shaw, who showed me how such communication was possible without a sacrifice in the clarity or integrity of ideas. If the reader obtains a better view of the forest of physics rather than of the trees, then I owe much to them.

I would like to thank the Aspen Center for Physics for its hospitality during the writing of this book.

Finally I want to thank my wife, Elaine, and my son, Mark, whose loving support made writing less a labor and more a creative challenge.

For my parents

contents

foreword

As a physicist I want to share the excitement of the recent discoveries of physics with other people—discoveries giving insights into the ultimate structure of matter, the origin and end of the universe, and the new quantum reality. In the last ten years physicists have learned more about the universe than in previous centuries—they have seen a new picture of reality requiring a conversion of our imaginations. The visible world is neither matter nor spirit but the invisible organization of energy.

This book is divided into three parts. The first part, "The Road to Quantum Reality," describes the development of the quantum theory of the atom. Grasping quantum reality requires changing from a reality that can be seen and felt to an instrumentally detected reality that can be perceived only intellectually. The world described by the quantum theory does not appeal to our immediate intuition as did the old classical physics. Quantum reality is rational but not visualizable.

Another way the old physics differs from quantum physics is the way the determinism of a clock differs from the contingency of a pinball machine. Albert Einstein, who never accepted the randomness at the foundation of reality implied by the quantum theory, expressed his objection by stating, "I cannot believe that God plays dice." Yet almost every physicist today believes He does. We will look at randomness in the hand of a dice-playing God and see what it implies about reality.

The second part of the book describes "The Voyage into Matter." Physicists, extending human consciousness into the farthest reaches of space and time, deep into the structure of matter, find that beyond the molecule and atom lies a new realm. The core of the atom is the nucleus. The same forces that bind the atomic nucleus together produce a new set of particles—forms of matter never before seen—called hadrons, and these in turn are made out of yet more fundamental particles called quarks. Physicists have voyaged into the realm of quarks and other quantum particles from which everything in the universe can be made. Here, at the smallest distances ever reached by our instruments, they have discovered the basic laws that unify the forces of nature.

Understanding the world of these elementary particles requires combining the quantum theory and Einstein's special relativity theory of space and time. The result of this combination which describes the creation and destruction of the quantum particles is called relativistic quantum field theory. It represents one of the great intellectual accomplishments of this century and realizes a radical new picture of the material world. Physicists have found the unified field theories they have been seeking for decades—theories which use complex and beautiful mathematical symmetries. The language of such physical theories is highly mathematical, and that has been an obstacle to many people in sharing in the excitement of these recent discoveries; but here we will not use mathematics.

Using these new unified field theories, physicists reconstruct the first few seconds of the big bang at the beginning of time when the universe was a swirling fireball of quarks and other quanta. Everything we know came out of that fireball. How our universe was born through a succession of broken symmetries and how it might end is described.

Finally, there is a short third part, "The Cosmic Code," which describes the nature of physical laws and how physicists find them. This part also contains some personal reflections on the meaning of the scientific enterprise—that through the activity of science and technology the discovered order of the universe, what I call the cosmic code, becomes a program for historical change. The modern world is a response to the challenging discoveries of the quantum and the cosmos—discoveries which continue to shape our future and to transform our idea of reality.

New York, New York
Aspen, Colorado 1981

PART I

The Road to Quantum Reality

The Lord may be subtle, but
He is not malicious.
—Albert Einstein

1. The Last Classical Physicist

Still there are moments when one feels free from one's own identification with human limitations and inadequacies. At such moments, one imagines that one stands on some spot of a small planet, gazing in amazement at the cold yet profoundly moving beauty of the eternal, the unfathomable: life and death flow into one, and there is neither evolution nor destiny; only being.

—Albert Einstein

As a young boy growing up in suburban Philadelphia, I had few heroes. Albert Einstein was one of them. Reading about Einstein in the newspapers and Sunday supplements, I learned he was working on a unified field theory, whatever that was. Before Einstein, scientists thought that space went on forever—that the universe was infinite. But what Einstein proposed, what really excited me, was the notion of the curvature of three-dimensional space, for that meant that the universe could be finite.

Imagine that you are in an airplane flying above the surface of our earth. If you fly long enough in a straight path in any direction, you return to your starting point, going around the world in a circle. The surface of our earth may be viewed as two-dimensional curved space, a finite surface that closes on itself without a boundary or edge. It's harder to visualize a three-dimensional curved space closing on itself in the same way, but we *can* imagine flying into the universe in any direction, maintaining a steady course, and eventually returning to our starting point. As in our round-the-world flight in the airplane we would never encounter a physical boundary, a stop sign that says the universe ends here. Einstein in his general theory of relativity proved that the three-dimensional space of our universe can curve around itself and be finite just like the curved surface of the earth.

My friends and baseball companions thought I was crazy when I explained this to them, but I felt confident and pleased because I had Einstein backing me up. Later I learned that Einstein, anticipating such appeals to his authority, once ironically remarked, "For rebelling against every form of authority Fate has punished me by making me an authority."

I never met Einstein. He had died by the time I went to Princeton University to major in physics. But I have spoken with his friends and collaborators, many of whom were refugees like himself. Einstein was present at the birth of twentieth-century physics. One might say he fathered it.

Twentieth-century physics grew out of the previous "classical" physics inspired by the work of Isaac Newton in the late seventeenth century. Newton discovered the laws of motion and gravitation and successfully applied them to describing the detailed motion of the planets and the moon. In the century following Newton's discoveries, a new interpretation of the universe emerged: determinism. According to determinism, the universe may be viewed as a great clockwork set in motion by a divine hand at the beginning of time and then left undisturbed. From its largest to its smallest motions the entire material creation moves in a way that can be predicted with absolute accuracy by the laws of Newton. Nothing is left to chance. The future is as precisely determined by the past as is the forward movement of a clock. Although our human minds could never in practice track the movement of all the parts of the great clockwork and thus know the future, we can imagine that an all-knowing mind of God can do this and see past and future time laid out like a mountain range.

This rigid determinism implied by Newton's laws promotes a sense of security about the place of humanity in the universe. All that happens—the tragedy and joy of human life—is already predetermined. The objective universe exists independently of human will and purpose. Nothing we do can alter it. The wheels of the great world clock turn as indifferent to human life as the silent motion of the stars. In a sense, eternity has already happened.

As strange as it seems today, complete determinism was the only conclusion that could be reasonably drawn from classical Newtonian physics. Even the great scientific advances of the nineteenth century—the theory of heat called thermodynamics, and the theory of light as an electromagnetic wave by the Scottish physicist, James Clerk Maxwell—were worked out within the framework of deterministic physics. These theories were among the last triumphs of classical physics. They are today still seen as major achievements, but the deterministic world view they supported fell. It fell not because of some new philosophy or ideology, but because by the end of the nineteenth century experimental physicists contacted the atomic structure of matter. What they found was that atomic units of matter behaved in random, uncontrollable ways which deterministic Newtonian physics could not account for. Theoretical physicists responded to these new experimental discoveries by inventing a new physical theory, the quantum theory, between 1900 and 1926.

When the earliest version of the quantum theory was formulated in 1900 it was not clear that a clean break with Newtonian physics was inevitable. Attempts were made between 1900 and 1926 to reconcile the quantum theory of atoms with deterministic physics. Physicists hoped that even the tiniest wheels of the great clockwork, the atoms, would obey Newton's deterministic laws. After 1926 it became clear that a radical break with Newtonian physics was required, and determinism fell.

Like Isaac Newton two centuries before him, Albert Einstein is a major transitional figure in the history of physics. Newton accomplished the transition begun by Galileo, from medieval scholastic physics to classical physics; Einstein pioneered the transition from Newtonian physics to the quantum theory of atoms and radiation, a new non-Newtonian physics. But the irony was that Einstein, who opened the route to the new quantum theory that shattered the deterministic world view, rejected the new quantum theory. He could not intellectually accept that the foundation of reality was governed by chance and randomness. Yet Einstein had led the tribe of physicists through a period of struggle into the promised land of the quantum theory, a theory which he could not see as giving a complete picture of physical reality. Einstein was the last classical physicist.

Why did Einstein reject the interpretation of the new quantum physics—the ultimate randomness of reality—when most of his fellow scientists accepted it? Any answer to this question cannot be simple. Einstein's rejection reflects not just his rational choice but also the roots of his personality and character formed during his childhood in Germany. By examining his childhood we find clues to his later persistent adherence to the classical world view.

Einstein was born in Ulm, Germany, on March 14, 1879, into a middle-class Swabian Jewish family. Shortly thereafter, his family moved to Munich, where Einstein's father started a small electrochemical business. Einstein was not an exceptional child and had a poor memory for words, often repeating the words of others softly with his lips. His mind played with spatial rather than linguistic associations; he built card towers of great height and loved jigsaw puzzles. When he was four his father gave him a magnetic compass. Seven decades later, in his "Autobiographical Notes" appearing in the volume *Albert Einstein: Philosopher-Scientist* he recalled the wonder that this compass inspired; it "did not at all fit into the nature of events which could find a place in the unconscious world of concepts. . . ."

Einstein's mother and father encouraged the young boy's curiosity. In a psychoanalytic study of Einstein's childhood, Erik Erikson called him "Albert, the victorious child." Something in Einstein's character and upbringing encouraged a profound sense of trust in the universe and life. That trust and the confidence it brings is the foundation of the autonomous mind living at the boundary of human knowledge.

His family had a liberal secular orientation. They were not especially intellectual but they respected learning and loved music. His parents, not being religiously observant, sent the young boy to a Catholic school, where he became involved with the ritual and symbolism of religion. This involvement was not to last. He wrote about his early emotional and intellectual odyssey from religion toward science when he was sixty-seven. These "Autobiographical Notes" display a simplicity and strength that characterizes his prose:

> Even when I was a fairly precocious young man the nothingness of the hopes and strivings which chases most men restlessly through life came to my consciousness with considerable vitality. Moreover, I soon discovered the cruelty of that chase, which in those years was much more carefully covered up by hypocrisy and glittering words than is the case today. By the mere existence of his stomach everyone was condemned to participate in that chase. Moreover, it was possible to satisfy the stomach by such participation, but not man in so far as he is a thinking and feeling being. As the first way out there was religion, which is implanted into every child by way of the traditional education-machine. Thus I came—despite the fact that I was the son of entirely irreligious (Jewish) parents—to a deep religiosity, which, however, found an abrupt ending at the age of 12. Through the reading of popular scientific books I soon reached the conviction that much in the stories of the Bible could not be true. The consequence was a positively fanatic [orgy of] freethinking coupled with the impression that youth is intentionally being deceived by the state through lies; it was a crushing impression. Suspicion against every kind of authority grew out of this experience, a skeptical attitude towards the convictions which were alive in any specific social environment—an attitude which has never again left me, even though later on, because of a better insight into the causal connections, it lost some of its original poignancy.
>
> It is quite clear to me that the religious paradise of youth, which was thus lost, was a first attempt to free myself from the chains of the "merely personal," from an existence which is dominated by wishes, hopes and primitive feelings. Out yonder

> there was this huge world, which exists independently of us human beings and which stands before us like a great, eternal riddle, at least partially accessible to our inspection and thinking. The contemplation of this world beckoned like a liberation, and I soon noticed that many a man whom I had learned to esteem and to admire had found inner freedom and security in devoted occupation with it. The mental grasp of this extrapersonal world within the frame of the given possibilities swam as highest aim half consciously and half unconsciously before my mind's eye. Similarly motivated men of the present and of the past, as well as the insights which they had achieved, were the friends which could not be lost. The road to this paradise was not as comfortable and alluring as the road to the religious paradise; but it has proved itself as trustworthy, and I have never regretted having chosen it.

What this passage reveals is a conversion from personal religion to the "cosmic religion" of science, an experience which changed him for the rest of his life. Einstein saw that the universe is governed by laws that can be known by us but that are independent of our thoughts and feelings. The existence of this cosmic code —the laws of material reality as confirmed by experience—is the bedrock faith that moves the natural scientist. The scientist sees in that code the eternal structure of reality, not as imposed by man or tradition but as written into the very substance of the universe. This recognition of the nature of the universe can come as a profound and moving experience to the young mind.

Many intellectual biographies of the turn of the century record a similar conversion. The symbols of religion and family are replaced by those from literary, political, or scientific culture. The formative event is the assertion of the individual's autonomy against parental, social, or religious authoritarianism. For Einstein this event took the form of liberating himself from a random existence "dominated by wishes, hopes and primitive feelings." He turned to the contemplation of the universe, a magnificent and orderly system that was, in his view, completely determined and independent of human will. The classical world view of reality fulfilled the needs of the young Einstein. The idea that reality is independent of how we question it may have been instilled in him then. This early commitment to classical determinism was to be the theme of his later opposition to the quantum theory, which maintains that fundamental atomic processes occur at random and that human intention influences the outcome of experiments.

When he was twelve Einstein received Euclid's geometry, "the holy geometry book," from his Uncle Jacob, and now Euclid became his Bible. Euclid's geometry appeals to reason, not authority or tradition. The new way of thinking attracted Einstein, and he became strongly antireligious and challenged the school's authoritarianism and discipline. No doubt the boy was a difficult student. He detested the military organization of German schools. He was rarely found in the company of children his own age, and once was even expelled from school by a teacher who said his mere presence in the classroom was sufficient to undermine the educational process.

When Einstein was fourteen, his father's business failed and the family moved to Italy. Albert did not at first join them but remained in Munich during 1894 attempting to finish school at the gymnasium. But he became a school dropout by the end of the year, joined the family in Italy, and spent most of the next year wandering in Italy, assuming his gymnasium teachers' recommendation would suffice to get him into a university. It did not, and he had to take an exam to enter Zurich Polytechnic Institute, which he failed. Then in the fall of 1895 he entered the Cantonal School of Argau, a Swiss preparatory school in the liberal Pestalozzi tradition to which he responded enthusiastically. Here he got his diploma, and in 1896 he entered the Zurich Polytechnic Institute to begin his education as a physicist.

Sometime in this year he first asked himself the question of what would happen if he could catch up to a light ray—actually move at the speed of light. The prevailing theory of light at that time—still valid today—was Maxwell's theory that light is a combination of electric and magnetic fields that move like a water wave through space. Einstein knew Maxwell's theory of light and the fact that it agreed with most experimental data. But if you could catch up to one of Maxwell's light waves the way a surfboard rider catches an ocean wave for a ride, then the light wave would not be moving relative to you but instead be standing still. The light wave would then be a standing wave of electric and magnetic fields that is not allowed if Maxwell's theory is right. So, he reasoned, there must be something wrong with the assumption that you can catch a light wave as you can catch a water wave. This idea was a seed from which the special theory of relativity grew nine years later. According to that theory, no material object can attain the speed of light. It is the speed limit for the universe.

In 1900, Einstein graduated from the university, but only by cramming for final exams. He detested the exams so much that he later commented that it had destroyed his motivation for scientific work for at least a year. He held various teaching jobs and tutored two young gymnasium students. Einstein went so far as to advise their father, a gymnasium teacher himself, to remove the boys from school, where their natural curiosity was being destroyed. He didn't last in that job.

Through a friend, he got a job at the patent office in Bern in 1902 while he worked on his doctorate. He earned his living examining patent applications and in his spare time worked on physics. This arrangement ideally suited him, for he never felt he ought to be paid to do theoretical physics research. In this modest way his career in physics began.

Theoretical physics at that time was dominated by the classical deterministic world view which had produced the great achievements of nineteenth-century physics—the theory of heat and Maxwell's electromagnetic theory. There was every reason to suppose it would continue. A major theoretical problem was how to deduce the laws of mechanical motion of electrically charged particles from the electromagnetic theory.

But experimental physicists had turned up some puzzles that did not have an explanation in terms of the prevailing theories. Radioactivity—the spontaneous emission of particles and rays from specific materials—had been observed. Perhaps the most puzzling observation of all was the sharp lines in the color spectrum of light emitted from different materials. No one had an explanation for that. These observations were like the first drops of rain in a storm that was soon to become a deluge sweeping away classical physics.

The puzzling experiments were indirectly revealing the properties and structure of matter down at the smallest distances beyond where anything could yet be directly seen. Today we know that the structure of matter at these small distances is atomic, but in Einstein's day some physicists still debated the existence of atoms. For over two millennia people had suspected the existence of atoms, but there had never been a way of proving their existence. In spite of all the indications, most especially from chemistry, that the atomic hypothesis—the hypothesis that all matter is made of atoms—was indeed correct, no one had devised a direct test to prove that atoms actually existed. Some leading scientists did not believe in atoms, including Ernst Mach, a philosopher-physicist. He was a positivist who maintained that all physical theory must come only from direct experimental experience, that all ideas that cannot be tested experimentally must be abandoned—the "seeing is believing" approach to physics. Mach did not believe in atoms because he had never seen one—and his strict viewpoint and rigorous thinking had a terrific impact on physics in general and upon Einstein in particular.

Max Planck was the physicist who brought forth the first crucial idea of the quantum theory in 1900, the same year Einstein graduated from the university. Previous to Planck's idea, most physicists conceived of the classical world of nature as a continuum: They thought of the forms of matter blending into one another in a smooth, continuous way. Various physical quantities like energy, momentum, and spin were continuous and could take on any value.

The basic idea of Planck's quantum hypothesis is that this continuous view of the world must be replaced by a discrete one. Because the discreteness of physical quantities is so very small, their discreteness is not perceptible to our senses. For example, if we look at a pile of wheat from a distance it appears to be a continuous smooth hill. But up close, we recognize the illusion and see that in fact it is made of tiny grains. The discrete grains are the quanta of the pile of wheat.

Another example of this "quantization" of continuous objects is the reproduction of photographs in newspapers. If you look closely at a newspaper photo it consists of lots of tiny dots; the image has been "quantized"—something you do not notice if you view the photo from afar.

Planck was struggling with the problem of black-body radiation. What is black-body radiation? Take a material object—a metal bar will do—and put it into a dark, light-tight room. The metal bar is the black body; that is, you cannot see it. If you heat the bar on a fire to a high temperature and return it to the dark room, it ceases to be black, instead glowing a dark red like a burning coal in a campfire. If you heat it to a still-higher temperature, the metal glows white hot. The light coming from the hot metal in a dark room has a distribution of colors which can be measured, resulting in what is called the black-body radiation curve.

Two teams of experimental physicists at the Physikalisch-Technische Reichsanstalt in Berlin made precise measurements of the black-body radiation curve. After fitting their empirical curve using ideas from the theory of heat, Planck tried to understand the physical basis for the new radiation law. Then with an incredible leap of intuition, Planck made the quantum hypothesis, which in his own words he described as "an act of sheer desperation." He supposed that the material of the black body consisted of "vibrating oscillators" (actually those were the atoms out of which the black body is made) whose energy exchange with the black-body radiation was quantized. Energy exchange was not continuous but discrete. Completely without precedent, this idea was one of the great leaps of the rational imagination, and Planck spent the remainder of his long life attempting to reconcile his radiation law with the continuous picture of nature.

Planck specified the amount of discreteness by a number h, later called Planck's constant. It specified, if you like, the size of a single grain in the pile of wheat. If Planck's constant could be set to zero, the grain reduced to zero size, then the continuous nature of the world would reappear. The experimental fact that Planck's constant h is not zero came to mean the world is in fact discrete. Planck, with the aid of his quantum hypothesis and some guesses, deduced the experimentally observed black-body radiation law. The Berlin experimentalists, in their report to the Prussian Academy on October 25, 1900, said the "formula, given by Herr M. Planck after our experiments had already been concluded . . . reproduces our observations within the limits of error." This was the beginning of the quantum theory. Einstein was twenty-one.

The world of theoretical physics that Einstein entered was dominated by the deterministic world view inspired by Newton's mechanics. Planck's work on the quantum broke with the idea of the continuum in nature, which was one of the main reasons for its neglect by physicists. Some puzzling experiments existed, but most physicists did not wish to give up Newton's laws to explain them. Scientific opinion was divided on the existence of atoms.

In 1905, the year he received his doctorate in Zurich, Einstein published three papers in volume 17 of *Annalen der Physik,* altering the course of scientific history. The volume is now a collector's item. Each of the three papers is a scientific masterpiece reflecting one of Einstein's three major interests: statistical mechanics, the quantum theory, and relativity. These papers began the physics revolution of the twentieth century. It would be decades before a new consensus on the nature of physical reality could be formed.

The first paper was on statistical mechanics, a theory of gases invented by James Clerk Maxwell, the Austrian physicist Ludwig Boltzmann and the American, J. Willard Gibbs. According to statistical mechanics, a gas like air consists of lots of molecules or atoms bouncing off each other in rapid random motion like a room filled with flying tennis balls. The tennis balls hit the walls, each other, and anything in the room. This model imitates the properties of a gas. But the atomic hypothesis that a gas actually consists of tiny atoms and molecules too small to see all flying around seems to be incapable of direct test.

It is hard to appreciate the atomic hypothesis because atoms are so small and there are so many of them. For example, in your last breath it is almost certain that you have inhaled at least one atom from the dying breath of Julius Caesar as he lamented, "Et tu, Brute." That is scientific trivia. But the fact is that a human breath contains about one million billion billion (10^{24}) atoms. Even if they mix with the entire atmosphere of the earth, the chances are high that you will inhale one of them.

We can't see or touch atoms; they are not a perceivable part of our world. Yet much of physics is based on the existence of atoms. Richard Feynman, one of the inventors of quantum electrodynamics, once wrote that if all of scientific knowledge were destroyed in some cataclysm except for one sentence which would be passed on to the future, it should be, ". . . all things are made of atoms—little particles that move around in perpetual motion, attracting each other when they are a little distance apart, but repelling upon being squeezed into one another."

The problem Einstein addressed was how to prove the existence of atoms. How could he do that when atoms were too small to be seen? Suppose you put a basketball into the room full of flying tennis balls. The big basketball gets bombarded from all sides by the tennis balls, and it begins to move in a random way. Assuming the randomness of the bombardment by the tennis balls, the features of the movements of the basketball can be determined. It jumps and bounces around because of the balls hitting it.

Einstein's paper made use of a similar idea to furnish the first convincing proof of the existence of atoms. He recognized that if you put into a gas or liquid relatively large grains of pollen—which could be seen under a powerful microscope—you could see them move around. The English botanist Robert Brown had observed this movement of pollen grains long before Einstein wrote his paper, but he had no explanation for his observation. Einstein explained that this Brownian movement of the pollen grains is due to atoms hitting the grains. The pollen grains are so small they get bounced and jiggled by the atoms hitting them just as would be a basketball being hit by tennis balls. Perrin, the French experimentalist, did some remarkable experiments that confirmed Einstein's quantitative predictions for the motion of the pollen grains. Many physicists then accepted the atomic hypothesis. Ostwald, the chemist, who didn't believe in atoms for reasons of his own, was converted to atomism by Einstein's analysis and Perrin's experiments. Ernst Mach, the strict positivist, was, however, never convinced of the existence of atoms, maintaining his "incorruptible skepticism" to his death. Physicists today recognize the paper of the patent examiner Einstein as proposing the first convincing test for the existence of atoms. That single paper alone would have made his scientific reputation.

The second bombshell paper of 1905 was Einstein's paper on the photoelectric effect. If a beam of light shines on a metal surface, electrically charged particles, electrons, are emitted by the metal, causing an electric current to flow. This is the photoelectric effect—light produces an electric current. The photoelectric effect is used in automatic elevator doors. A beam of light crossing the elevator door hits a metal surface, causing an electric current to flow. If the current flows, the door will close. But if the beam of light is interrupted by a person walking through the door, the current stops and the door stays open.

In 1905, little was known about the photoelectric effect. It is characteristic of Einstein's genius that he was able to see in this obscure physical effect a deep clue about the nature of light and physical reality. The creative movement in science moves from the specific—like the photoelectric effect—to the general—the nature of light. In a grain of sand one may see the universe.

Einstein, in his paper on the photoelectric effect, used Planck's quantum hypothesis. He went beyond Planck to make the radical assumption that light itself was quantized into particles. Most physicists, including Planck, thought that light was a wavelike phenomenon in accord with the view of nature as a continuum. Einstein's hypothesis implied that actually light was a rain of particles consisting of the light quanta later called photons—little packets of definite energy. Using his idea of light quanta, Einstein deduced an equation to describe the photoelectric effect.

Of the three 1905 papers, Einstein referred only to the paper on the photoelectric effect as "truly revolutionary," and indeed it was. One thing physicists had thought they understood was light; they understood it as a continuous electromagnetic wave. Einstein's work seemed to deny this, to claim instead that light was a particle. This is one reason why other physicists resisted his revolutionary idea. Another reason was that, unlike Planck's formula for black-body radiation, which was immediately checked experimentally, there was simply no way to confirm Einstein's photoelectric equation experimentally—and there wouldn't be until 1915. His introduction of the light quantum seemed gratuitous.

Einstein stood alone for more than a decade on the question of energy quantization of light. When he was recommended for membership in the Prussian Academy of Sciences in 1913, the letter read, "In sum, one can hardly say that there is not one among the great problems, in which modern physics is so rich, to which Einstein has not made a remarkable contribution. That he may have missed the target in his speculations, as, for example, in his hypothesis of the light quanta, cannot really be held too much against him, for it is not possible to introduce really new ideas even in the exact sciences without taking a risk." Millikan, the American experimentalist, spent years working on the photoelectric effect, devising precise measurements to test Einstein's photoelectric equation. In 1915 he said, "Despite . . . the apparent complete success of the Einstein equation, the physical theory of which it was designed to be the symbolic expression is found so untenable that Einstein himself, I believe, no longer holds to it." Einstein held to it. But it was clear that even after his photoelectric equation was experimentally confirmed, other physicists resisted the idea that light is a particle. The "truly revolutionary" idea of the photon, the light particle, needed further experimental confirmation before it could be accepted.

The final confirmation of the photon came in 1923–24. Assuming that light consisted of true particles that had a definite energy and directed momentum like little bullets, Compton, one of the first American atomic physicists, and Debye, a Dutch physicist, independently made theoretical predictions for the scattering of photons from another particle, the electron. Compton performed the scattering experiments, and the predictions based on the light particle assumption were confirmed. Opposition to the photon concept fell rapidly after that. Einstein's Nobel Prize was for his light quantum hypothesis, not for his greatest work, the relativity theory.

Einstein's third 1905 article was on the special theory of relativity. This article changed forever the way we think about space and time. Max Planck said in 1910 of this paper, "If [it] . . . should prove to be correct, as I expect it will, he will be considered the Copernicus of the twentieth century." Planck was right.

The special theory of relativity—as the topic of his 1905 paper was later called—dealt with space and time concepts that philosophers and scientists had devoted much thought to over the ages. Some thought that space was a substance—the ether—which pervaded everything. Others evoked images of the flow of time like a river or sand falling in an hourglass. While such images appeal to our feelings, they have little to do with the concept of time in physics. Understanding space and time in physics requires that we distinguish our subjective experience of space and time from what we can actually measure about them. Einstein said it very simply: Space is what we measure with a measuring rod and time is what we measure with a clock. The clarity of these definitions reveals a mind intent on great purpose.

Armed with these definitions, Einstein asked how the measurement of space and time changes between two observers moving at a constant velocity relative to one another. Suppose one observer is riding on a moving train with his measuring rod and clock and his friend is on the station platform with his rod and clock. The person on the train measures the length of the window on the side of his car. Likewise, the person on the platform measures the length of the same window as it moves by. How do the measurements of the two observers compare? Naively, we would think they must agree—after all, it is the same window that is being measured. But this is incorrect, as Einstein showed by a careful analysis of the measurement process. The person standing on the platform with his measuring rod must "see" the window moving past him. In other words, light which bears information about the length of the moving window must be transmitted to the person standing on the platform, otherwise it can't be measured at all. The properties of light have entered our comparison of the two measurements, and we must first examine what light does.

Even before Einstein, physicists knew the speed of light was finite but very fast, about 180,000 miles per second. But Einstein thought there was something special about the speed of light—that the speed of light is an absolute constant. No matter how fast you move, the speed of light is always the same—you can never catch up to a light ray. To appreciate how odd this really is, imagine that a gun fires a bullet at some high speed. But the speed of a bullet is not an absolute constant, so that if we take off after the bullet in a rocket we can catch up to it and it appears to be at rest. There is no absolute meaning to the speed of the bullet because it is always relative to our speed. But not so with light; its speed is absolute—always the same, completely independent of our own velocity. That is the odd property of light that makes its speed qualitatively different from the speed of anything else.

The assumption of the absolute constancy of the speed of light was the second postulate of the special theory of relativity. The first postulate Einstein made was that it is impossible to determine absolute uniform motion. Uniform motion proceeds in a fixed direction at a constant speed—basically coasting. Einstein's postulate is that you cannot determine if you are coasting unless you compare your motion relative to another object. The two observers, one on the train, the other on the platform, illustrate this postulate. For the person on the platform it is the train that is moving. But the person in the train can just as well suppose he is standing still and the platform and the whole earth with it are moving past him. Uniform motion is only relative—you can only say you are moving relative to something else.

From these two postulates, the constancy of the speed of light and the relativity of motion, the entire logical structure of special relativity followed. But, as Paul Ehrenfest, a physicist and a friend of Einstein's, emphasized, there is implicit a third postulate which states that the first two are not in contradiction. Superficially, it seems that they are. All uniform motions are relative to one another, says one postulate. Except the motion of light, which is absolute, says the other postulate. It is the interplay between the relativity of motion for all material objects and the absoluteness of the speed of light which is at the root of all the unfamiliar features of the world according to special relativity.

Using these postulates, Einstein mathematically deduced the laws that related space and time measurements made by one observer to the same measurements made by another observer moving uniformly relative to the first. He showed that the person on the platform would actually find the length of the window on the moving train is shorter than the person on the train. As the train speeds up, the length of the window would be measured to be shorter and shorter by the person on the platform, until, as the imaginary train approached the speed of light, the length of the window would shrink to zero. Because in our familiar world the speed of most objects, like real trains, is so small compared to the speed of light, we never see such length contractions, which become dramatic only at speeds near that of light.

Einstein's theory of relativity linked space and time. Einstein showed that a moving clock marked time more slowly than one at rest. For the person on the platform, the watch on the wrist of the train's passenger actually moves more slowly—time slows down. If the train was moving near light velocity, time changes would actually slow down to close to zero. Likewise, the person on the train will see the watch of the person on the platform move more slowly. Absolute time is abolished. Time is measured differently for persons moving relative to one another.

It seems as if the relativity of time poses a paradox—for how can both the passenger on the train and the person on the platform *both* see each other's watches slow down? What happens if now these people meet and compare the time; whose watch has really slowed down? To emphasize this paradox (often called the twin paradox), imagine twins who each set their watches before one of them gets on the train. The train speeds up to nearly the speed of light—at which point each twin will see the other's watch running slower—and then the train slows down and returns to the station. Which twin is older? From the point of view of the twin on the platform, the one on the train has made a round-trip journey, while for the twin on the train, the twin on the platform is the one who has made the round trip. It seems as if the motion of each twin is simply relative to the other's motion. But there is in fact an asymmetry in the motion of the twins, and that is the clue for resolving the paradox. While the train is speeding up it is no longer in uniform motion but is accelerating, and later in the trip it is decelerating. The twin on the platform never experiences such acceleration and deceleration so there is an absolute distinction between the twins' motion. Einstein's special theory of relativity applies only to uniform motions, and the motion of the train is not always uniform. By using Einstein's general theory of relativity, which applies to nonuniform motions like that of the train, one can demonstrate that the twin on the train has actually aged less.

The relativity of space and time disturbs us because it contradicts our intuition. In everyday experience, space and time do not appear to shrink. We might want to think that these odd effects of space and time are merely a mathematical fiction. The French mathematician Poincaré independently discovered the same space-time transformation laws in 1905, but he thought of them as postulates, without physical significance. Einstein was the first to understand the physical implications of those laws; for this reason he is considered the inventor of relativity. He took the physics seriously: clocks really slow down when they move.

One way you might experience the space-time of special relativity, not conceptually but physically, is to imagine you are eleven million miles tall. It takes light about a minute to travel eleven million miles. If you decide to wriggle your toes—assuming nerve impulses could be speeded up to light speed—it would take a minute for the signal to get to your toes and still another minute to return to tell your brain the toes have indeed wriggled. You would feel as if you were in a slow-motion picture with a body made of elastic rubber. If you started walking, the upper part of your leg would move up long before the foot would lift, because the nerve impulses would get there first and to the foot only a half-minute later. Because the speed of light is finite, you could not lift your leg all at once in a coordinated manner—you would simply be unable to signal your foot, knee, and thigh to move together simultaneously. No signal moves faster than light; nothing moves instantaneously.

Or imagine two ordinary-sized people, one on earth, and the other on a spaceship moving at nearly the speed of light. Both have front-row seats to watch a performance of an eleven-million-mile-tall dancer who moves across the solar system as if it were a stage. It is a marvelous performance; and later they discuss it but cannot agree on what they saw. The viewer on the spaceship says the dancer first moved her arm and then her foot, but the viewer on earth saw these events in reverse order. Even if they try to analyze the motion of the dancer taking into account the finite speed of light and the motion of the spaceship and the earth, they cannot agree. The reason is that the second postulate of special relativity—that the speed of light is an absolute constant—denies the existence of a universal time for all observers. Even the order of events in time can be different for observers moving relative to one another; there is no absolute meaning to such time orderings. The consequences of special relativity seem paradoxical compared to our everyday experience. The unfamiliar world of special relativity becomes apparent only when speeds approach that of light; the speeds we encounter in everyday life are not near that. But special relativity is a logically consistent and coherent theory; there are no paradoxes.

Einstein wrote a final, fourth short paper in 1905, the full consequences of which were not developed until 1907. By an analysis of the energy of motion E, of a relativistic particle of mass m, he

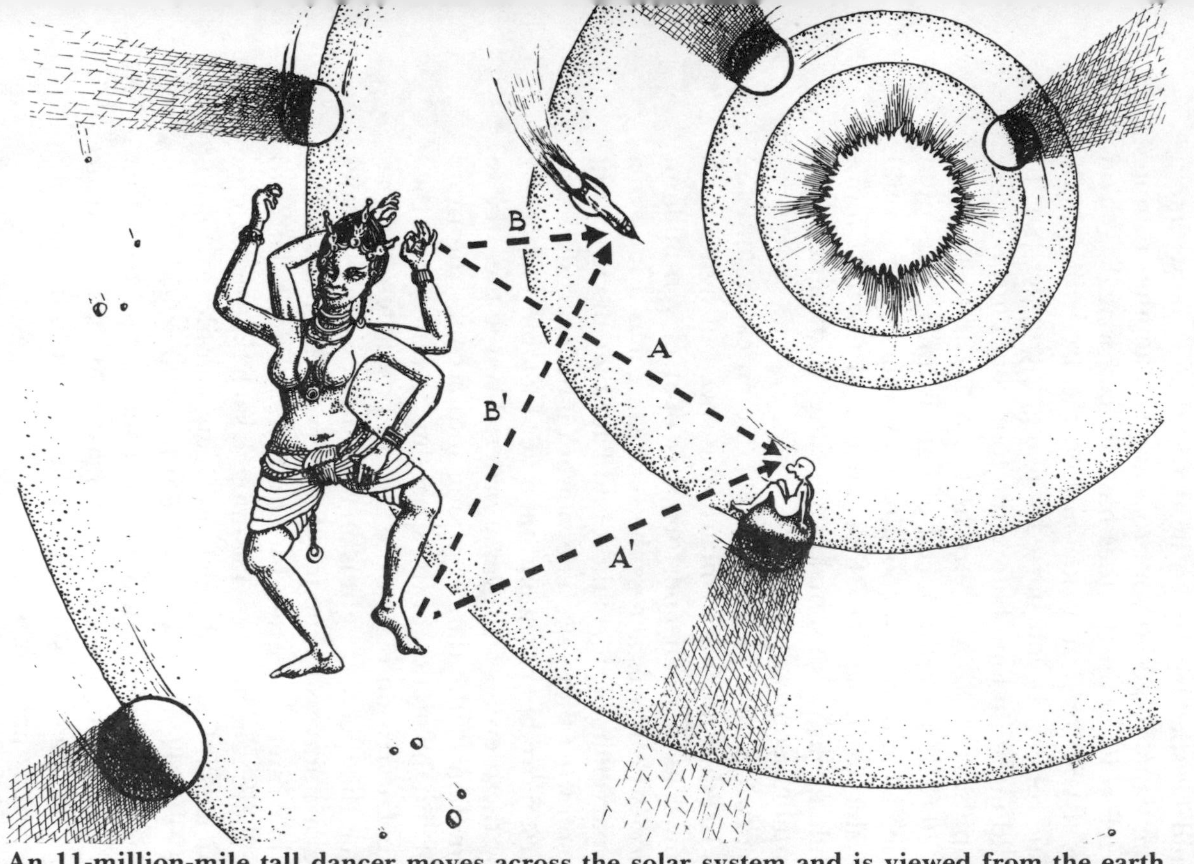

An 11-million-mile tall dancer moves across the solar system and is viewed from the earth and from a spaceship moving at nearly light velocity relative to the earth. The observers on earth and on the spaceship cannot agree whether the dancer first moved her hand or her foot. Even after taking into account their relative motion and the finite speed of light they cannot agree which event "really" took place first. Unlike the Newtonian concept of time there is no universal time according to special relativity theory.

showed that the particle had an energy given by $E = mc^2$. The constant c is the speed of light.

Before Einstein, physicists thought of energy and mass as distinct. This seems evident from our experience. What does the energy we expend by lifting a stone have to do with the mass of the stone? Mass conveys the impression of a material presence, while energy does not.

Mass and energy were also quantities that seemed to be separately conserved. In the nineteenth century, physicists discovered the law of conservation of energy—it can neither be created nor destroyed. If you lift a stone, energy has been expended but not lost. The stone has potential energy that is released if the stone is dropped. There was also a separate conservation law for mass—mass could neither be created nor destroyed. If a stone is broken up, the pieces have the same total mass as the original stone. The distinction of energy and mass and their separate conservation was deeply embedded in the thinking of physicists in 1905, because it had enormous experimental support. With that background of thought, the novelty of Einstein's insight may be contrasted.

Einstein discovered that the postulates of relativity theory implied that the distinction between energy and mass and the notion of their separate conservation had to be abandoned. This shattering discovery is what is summarized in his equation $E = mc^2$. Mass and energy are simply different manifestations of the same thing. All the mass you see about you is a form of bound energy. If even a small part of this bound energy were ever released, the result would be a catastrophic explosion like that of a nuclear bomb. Of course, the matter around us is not about to convert itself into energy—it takes very special physical conditions to accomplish that. But at the beginning of time during the big bang that created the universe, mass and energy were freely converting into one another. Today energy and matter only appear distinct, and someday in the far future the matter we see about us may again be freely converting into energy.

How well tested is the theory of special relativity? Today there is a whole technology that depends on the correctness of the theory—practical devices that simply would not work if special relativity were wrong. The electron microscope is one such device. The focusing of the electron microscope takes into account effects of relativity theory. The principles of relativity theory are also incorporated into the design of klystrons, electronic tubes that supply microwave power to radar systems. Perhaps the best evidence that special relativity theory works is the operation of the huge particle accelerators that accelerate subatomic particles like electrons and protons nearly to light velocity. The two-mile-long electron accelerator near Stanford University in California accelerates electrons until their mass increases as predicted by relativity by a factor of forty thousand at the end of their two-mile journey.

One of the oddest predictions of relativity theory is the slowing down of moving clocks. Interestingly, this is one of the most precisely tested predictions of the theory. While we can't accelerate real clocks up to the speed of light, there does exist a tiny subatomic particle, the muon, that behaves just like a tiny clock. After a fraction of a second, the muon disintegrates into other particles. The time it takes the muon to disintegrate may be thought of as a single tick of this tiny clock. By comparing the lifetime of a muon at rest with one that is rapidly moving, we can know how much this tiny clock has slowed down. This was done at CERN, a nuclear laboratory near Geneva, Switzerland, by putting the rapidly moving muons into a storage ring and precisely measuring their lifetime. The observed increase in their lifetime was a precise confirmation of the slowing of moving clocks predicted by special relativity.

These and many other tests confirm the correctness of the early work of Einstein. The young Einstein was a bohemian and a rebel who identified himself with the highest and best in human thought. During his period of intense creativity from 1905 to 1925 he seemed to have a hotline to "the old One"—his term for the Creator or Intelligence of nature. His gift was an ability to go to the heart of the matter with simple and compelling arguments. Separate from the community of physicists but in touch with the perennial problems of his science, Einstein realized a new vision of the universe.

Einstein's papers of 1905 and Planck's paper of 1900 ushered in the physics of the twentieth century. They transformed the physics that went before. Planck's idea of the quantum, further developed by Einstein as a photon, the particle of light, implied that the continuous view of nature could not be maintained. Matter was shown to be composed of discrete atoms. The ideas of space and time held since the age of Newton were overthrown. Yet in spite of these advances, the idea of determinism—that every detail of the universe was subject to physical law—remained entrenched in Einstein and his entire generation of physicists. Nothing in these discoveries challenged determinism.

Einstein's great strength lay not in mathematical technique but in a depth of understanding and a steadfast commitment to principles. That commitment to the principles of classical physics and determinism now moved him from his work on special relativity toward his greatest work, the general theory of relativity.

2. Inventing General Relativity

But the creative principle resides in mathematics. In a certain sense, therefore, I hold it true that pure thought can grasp reality, as the ancients dreamed.
—Albert Einstein

Recognition for Einstein began with the papers of 1905—the test for the existence of atoms, the introduction of the photon as the particle of light, and the special theory of relativity. In the fall of 1909 he left his job in the patent office to accept a faculty position at the University of Zurich followed by positions at the German University in Prague and then at the Zurich Polytechnic. In 1913, Max Planck visited Einstein in Zurich and offered him the best position in Europe for a theoretical physicist, the directorship of the Institute of Physics at the Kaiser Wilhelm Institute in Berlin. Einstein accepted. He was also offered a chair at the Prussian Academy and a professorship at the University of Berlin. In spite of his resistance to returning to Germany and to the academic world he disliked, this position offered him the opportunity of working with the greatest physicists of his time, including Planck.

Associating with these physicists was one of the influential experiences of his life. In Berlin, Einstein contributed to the theory of specific heats and gave a new derivation of Planck's black body radiation law. In this latter work he used his new idea of light particles, photons, and introduced the concept of stimulated light emission, the principle on which the modern laser operates.

Einstein completed his greatest work—the general theory of relativity—in Berlin during 1915–1916. This theory extended the concepts of space and time already introduced in his earlier work. Previously, in the special theory of relativity, Einstein had discovered the laws relating space and time measurements between two uniformly moving observers (such as the person on the train and the one on the station platform). A uniform motion is one that proceeds at constant speed and fixed direction. By contrast, a nonuniform motion is one in which the speed is changing (the train accelerates or slows down) or changes direction (the train goes around a curve). But in order to treat such nonuniform motion Einstein realized that he had to go beyond the postulates of special relativity.

Suppose that instead of the train we used to illustrate special relativity we are in a spaceship far from the earth. When the rocket engines are turned on, the spaceship begins to move, slowly at first, then faster and faster. Since the speed is increasing, this is an acceleration—a nonuniform motion of the spaceship. Inside the spaceship we experience this acceleration as a force pressing us to the floor. As long as the rocket engines accelerate the ship we continue to feel the force.

Remarkably, this force, which we know is due to the accelerating spaceship, cannot be distinguished from gravity. If we drop stones of different masses inside the accelerating spaceship, they will fall to the floor at the same rate—just as they do if we drop them here on earth. The moment we release the stones, they cease to be accelerated by the spaceship—they are in free fall—and we may think it is the floor of the spaceship that rushes up to contact them.

This illustrates the first main idea of general relativity—that it is impossible to distinguish the effect of gravity from a nonuniform motion (like that of the accelerating spaceship). Inside the spaceship we feel real gravity. If we didn't know we were in outer space traveling in a spaceship, then we could not determine that the effect of "gravity" we feel was due to the accelerating movement of the entire ship. The fact that we cannot physically distin-

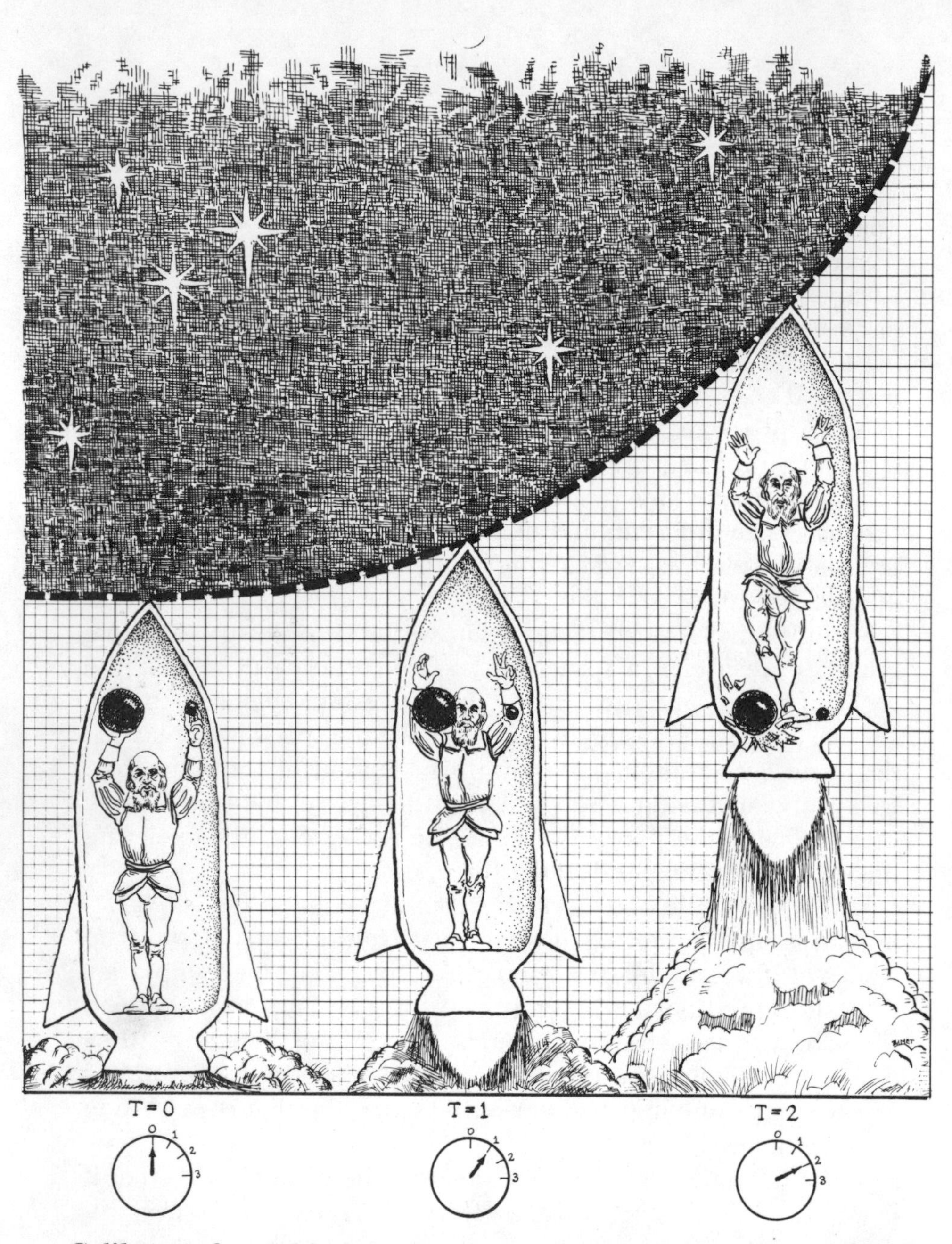

Galileo performs his legendary experiment, not from the leaning tower at Pisa, but inside an accelerating spaceship. He releases two objects of different mass which seem to him to be falling exactly as they would on earth. But notice that the two balls are in fact not accelerating—they are in "free fall" and are not subject to any external forces. It is the floor of the accelerating spaceship that rushes up to meet them. This illustrates the equivalence of accelerated motion and gravity—the first postulate of general relativity theory.

guish a nonuniform motion like an acceleration from gravity is called the principle of equivalence—the equivalence of nonuniform motion and gravity.

Einstein recorded the creative moment, "the happiest thought of my life," when he saw how all this fit together:

> When, in the year 1907, I was working on a summary essay concerning the special theory of relativity for the Yearbook for Radioactivity and Electronics, I tried to modify Newton's theory of gravitation in such a way that it would fit into the theory. Attempts in this direction showed the possibility of carrying out this enterprise, but they did not satisfy me because they had to be supported by hypotheses without physical basis. At that point there came to me the happiest thought of my life in the following form:
>
> Just as in the case where an electric field is produced by electromagnetic induction, the gravitational field similarly has a relative existence. *Thus, for an observer in free fall from the roof of a house there exists, during his fall, no gravitational field* [Einstein's italics], at least not in his immediate vicinity. If the observer releases any objects, they will remain relative to him in a state of rest, or in a state of uniform motion, independent of their particular chemical and physical nature. (In this consideration one must naturally neglect air resistance.) The observer is therefore justified in considering his state as one of "rest."
>
> The extraordinarily curious empirical law that all bodies in the same gravitational field fall with the same acceleration immediately took on, through this consideration, a deep physical meaning. For if there is even one thing which falls differently in a gravitational field than do the others, the observer would discern by means of it that he is falling in it. But if such a thing does not exist—as experience has confirmed with great precision—the observer lacks any objective ground to consider himself as falling in a gravitational field. Rather, he has the right to consider his state as that of rest and his surroundings (with respect to gravitation) as field-free.
>
> The fact, known from experience, that acceleration in free fall is independent of the material is therefore a mighty argument that the postulate of relativity is to be extended to coor-

dinate systems that are moving nonuniformly relative to one another.

Einstein grasped that the effect of gravity was equivalent to a nonuniform motion. Standing on the earth, we feel gravity pulling us to the ground. If we drop a stone, it falls. But if we fall from the roof of a house, there is no gravity. If we now drop a stone during our fall from the roof, it floats in front of us. It is like being in a spaceship that is not accelerating—we are in free fall and there is no gravity. Astronauts experience a gravity-free environment when the rocket engines are shut off and acceleration ceases.

If we fall or drop a stone in our room on board the accelerating spacecraft, we may perceive that it is the floor that is accelerating up. On the earth it is not obvious that the effect of gravity we experience is equivalent to the ground accelerating up. But it is—gravity is precisely equivalent to nonuniform motion.

In the general theory of relativity, Einstein found the laws relating space and time measurements carried out by two observers moving nonuniformly (one observer in an accelerating spaceship, for instance, and the other floating in gravity-free space). Considering these laws took Einstein into the mathematical discipline of Riemannian geometry—the geometry of curved space. Here Einstein solicited the help of a mathematician friend and former classmate, Marcel Grossman. However, even before Einstein undertook these mathematical investigations to generalize the relativity principle, he already intuited the result. As he remarked, "I first learned of the work of Riemann at a time when the basic principle of the general theory of relativity had already been clearly conceived." The creation of general relativity offers an example of a physicist turning to an existing mathematical discipline to find the right language to express his intuitions.

Why did Einstein need to consider curved space to describe gravity? The curvature of three-dimensional space (four dimensions if we include time) is hard for our minds to grasp. Let us first imagine a space that has only two dimensions, like a tremendous sheet of paper extending infinitely in all directions. The inhabitants of this sheet of paper are flat shadows—they have only two dimensions—and don't know anything about the third dimension. On their sheet of paper they can carry out geometrical measurements. The world they live in has Euclidean geometry—it is flat. If they measure the sum of the angles on the interior of a triangle drawn on the paper they would get 180°, in accord with a theorem of Euclidean geometry. Two parallel lines drawn on this paper would never meet if extended—another feature of flat space.

Now we move our two-dimensional shadow creatures to a new world, the surface of a large sphere. While we, as three-dimensional creatures, can see their sphere as a three-dimensional object in space, the shadow creatures can know only the surface of the sphere—a two-dimensional space similar to the sheet of paper they just left. What is interesting is how our shadow friends come to learn of the difference between the two-dimensional surface of the sphere and that of the sheet of paper. At first, the shadow creatures are quite happy in this new world because it seems so similar to the world they left. If they draw small triangles and measure the angles on the inside as best they can, the angles add up to 180°. Locally their new world is Euclidean and flat. Then the shadow creatures make a technological breakthrough—they discover a kind of laser light that can send out a straight-line beam along the spherical surface of their world for thousands of miles. The first thing they notice is that if two beams of light are sent out in parallel directions, they start coming together after traveling a thousand miles. No amount of adjustment corrects for this. Some shadow creatures argue that light beams don't move in straight lines in the new world. Others insist that a light beam is by definition a straight line: A light beam continues to travel along the shortest path; any other path will be longer. They realize there is nothing wrong with the light beam—rather, the space they move in is curved and not flat. If big triangles are made with these light beams the sum of the angles is now larger than 180°. Clearly, the space is not Euclidean. Eventually the shadow crea-

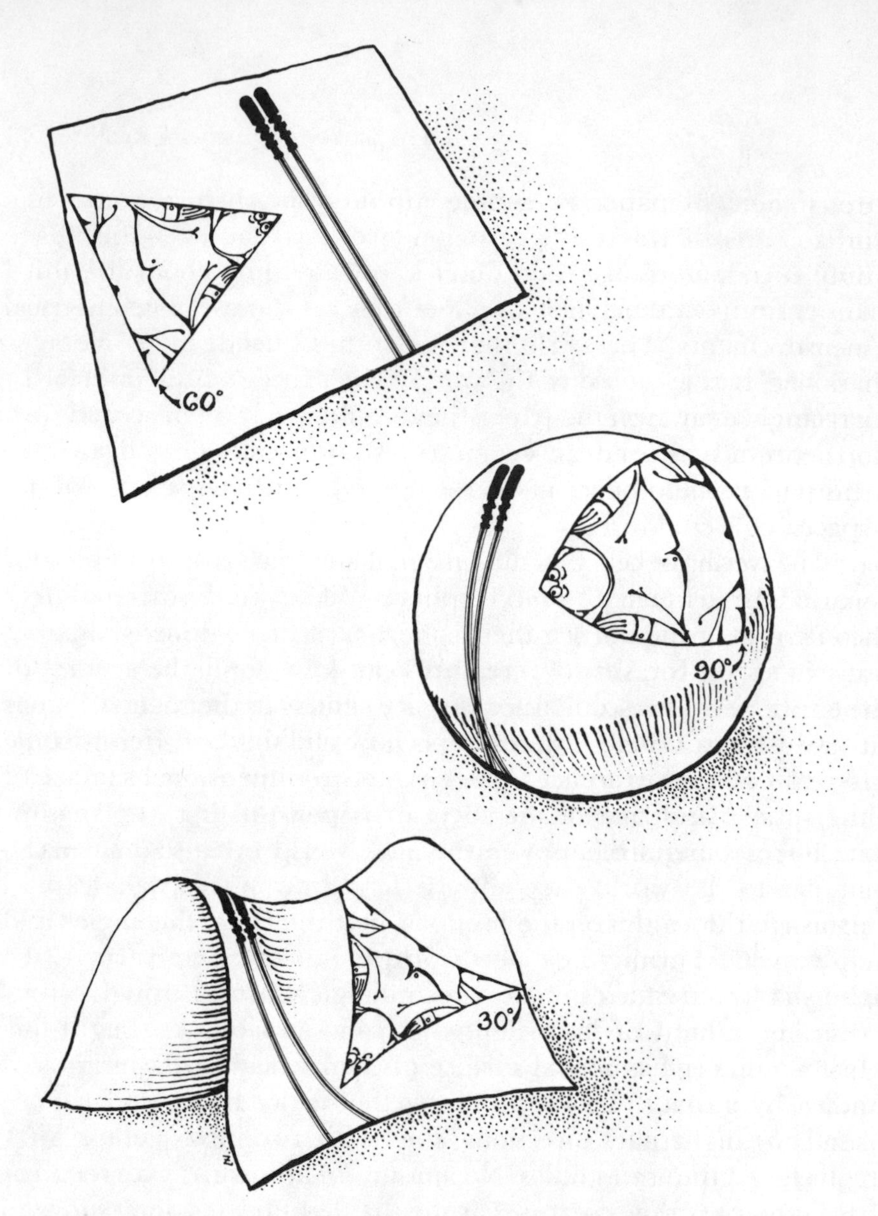

A two-dimensional scientist explores three different two-dimensional geometrical surfaces. At the top is the open, flat Euclidean geometry for which the sum of the angles of a triangle is 180° and parallel laser beams never meet. In the middle is the closed surface of a sphere—a non-Euclidean space—and the angles can add to more than 180° and "parallel" laser beams must cross. At the bottom is the open, hyperbolic surface geometry, also a non-Euclidean space, for which the angles of a triangle are less than 180° and "parallel" laser beams diverge. The space of our three-dimensional universe can be similarly classified as either flat, spherical, or hyperbolic. It is a difficult experimental problem, as not yet settled, to determine which of these three geometries is actually realized for our universe.

tures invent Riemannian geometry to describe their new curved world.

Our own story is like that of our shadow friends, except that it takes place in three, not two, dimensions of space. We may live in a world that is a curved three-dimensional space. Just as the shadow creatures could not visualize the curved two-dimensional surface of their new world, we cannot visualize a three-dimensional curved space. But, like them, we can use experiments with the laser beams to find out whether our three-dimensional world is in fact curved. Most physicists would bet that if you sent out two parallel laser light beams across intergalactic space they would not remain parallel. They would either diverge or come together. If they diverged, the universe is said to be "open"—space is curved but it goes on forever. If the light beams eventually converged, the universe is "closed"—the three-dimensional analogue of the surface of a sphere. Which of these possibilities corresponds to the real universe is for experimental astronomers to decide. In either case, the space of our universe is non-Euclidean; it is not flat. The geometry of that space is described by Riemannian geometry.

But what does this curvature of space have to do with gravity and nonuniform motions? Once we decide to define a straight line by the path of a light ray we can easily see this relation.

Because a light ray has energy, Einstein's mass-energy equivalence implies it has an effective mass. Everything massive is attracted by gravity. This means if we shoot a light beam near a planet the light path will bend a little toward that planet. We might be tempted to say that the bending of the light path means that light paths really aren't straight lines any more. We would be like those shadow creatures who could not accept that light beams weren't parallel any more and blamed it on the light itself. Actually, the curvature of space—the very geometry of their world—was responsible. Likewise, we could blame the bending of light around a planet on "gravity," a mysterious force. But Einstein saw that gravity was a superfluous concept—there isn't any "gravitational force." What actually happens is that the mass of a planet

—or any mass—curves the space near it, altering its geometry. Light always moves in a straight line—but a straight line as defined in a curved space. Einstein dispensed with the notion of gravity in favor of the geometry of curved space. In effect, he discovered that gravity *is* geometry. That is the central conclusion of general relativity.

We may summarize the main ideas of general relativity as follows. First, we recognize the equivalence principle—that gravity and a nonuniform motion are indistinguishable. Second, as a separate idea, we must recognize that determining the geometry of space is an experimental problem. By shooting laser light beams around we can map out the curved geometry of our space. These two ideas, the equivalence principle and the curvature of space, can be combined if we recognize that the path of light—which we use to determine the curved geometry of space—is subject to the influence of gravity. The nonuniform motion of a light beam—its bending in space—is equivalent to the effect of gravity in that region of space. But rather than thinking that a light path "bends" in the presence of "gravity," we should instead realize that "gravity" is really manifested as curved space and light beams are moving along the shortest path in that curved space. Gravity is the curvature of space.

Einstein in his article on general relativity theory derived a set of equations that specified the curved geometry of space—equivalent to gravity—produced by the presence of matter, like the sun or a planet. These equations precisely determine how space gets curved due to the presence of matter. The old idea—going back to Newton—was that matter, like the earth, produces a gravitational field that attracts other matter to it. This idea is now replaced by Einstein's idea that matter changes the geometry of space from flat to curved in its vicinity.

Einstein proposed three experimental tests for the general theory of relativity—that gravity is the curvature of space and time. These are: (1) a slight bending of light in the gravitational field of the sun, (2) a small shift in the orbit of the planet Mercury, and (3) clocks should run slower in a gravity field.

The first test of general relativity is the bending of light around the edge of the sun. Today scientists perform this experiment using radio interferometers, devices which can precisely measure the position of distant radio sources like certain galaxies and stars as they pass behind the sun. But when Einstein proposed this experiment in 1916, there were no radio telescopes. Arthur Eddington, a British astronomer and member of the Royal Society, heard of Einstein's new theory and wanted to check it by observing a total eclipse of the sun which was to take place on May 29, 1919, in the Southern hemisphere. With the First World War raging, there was no hope for the Royal Society to obtain scarce funds to outfit a solar expedition. But Eddington was a pacifist and an embarrassment to his government; he got his £5,000 probably to get him out of England. The eclipse was observed in Sobral, Brazil, and also at Principe, an island off the coast of West Africa.

During a total eclipse, the field of stars very near the eclipsed sun becomes visible in the darkness, and they can be photographed. The light from those distant stars behind the sun is on a path passing very close to the edge of the sun and hence, according to Einstein, ought to bend in the curved space around the sun. This bending can be revealed if this photograph is compared with a second photograph of the same group of stars taken at night six months later when the sun is not near the light path of the stars. The comparison shows there is a shift in the relative positions of the stars in the two photographs caused by the bending of light in the curved space-time around the sun. In 1919 the Royal Society announced the result that the positions of stars seen during the eclipse both at Sobral and Principe agreed with Einstein's prediction. After two hundred years, Newton's law of gravitation was overthrown, and Einstein's public notoriety began.

It was not easy for the Royal Society to get the results of this crucial experiment to Einstein in Berlin, because the First World War was just over. The telegram, sent from London, first reached the physicist Hendrik Lorentz, in Holland, which was a neutral country. Lorentz sent it on to Einstein in Berlin. A student of

Einstein's was in his office, and Einstein interrupted his discussion with her to hand her the telegram from the windowsill with the words, "This may interest you." When she read the statement that the British solar expedition had confirmed Einstein's theory, she exclaimed that this was a very important message. But Einstein was not excited and said, "I knew that the theory was correct. Did you doubt it?" The student protested; what would Einstein have thought if the result of the experiment had not confirmed his theory of general relativity? Einstein responded, "Then I would have to be sorry for dear God. The theory is correct." The hotline to "the old One" was still open.

This classic test of relativity was done long ago. But only in the last decade have a number of new tests been devised that very precisely test general relativity. The technology simply didn't exist ten years ago.

Irwin Shapiro and his collaborators at MIT have devised a beautiful test of general relativity. Using a powerful radar beam and computer signal processors, they bounce radar off a planet such as Mercury or Venus just before it is eclipsed by passing behind the sun as seen from the earth. When the planet is eclipsed, no radar beam returns, but just before eclipse it is possible to measure the amount of time it takes for the radar signal (which is the same as a light beam) to leave the earth, reflect from the distant planet, and return to earth. According to general relativity, a light beam has to bend slightly as it passes very near the edge of the sun because of the space curvature. This increases the round-trip time for the light beam over what it would be if the beam did not graze the edge of the sun. As the planet approaches the edge of the sun as seen from earth, it takes a longer time for the radar signal to return, and general relativity has a precise prediction for this delay. Within small experimental errors the prediction is confirmed.

The advent of satellite technology and the exploration of the solar system by unmanned space probes opened new ways of testing general relativity. There is now an orbiter around the planet Mars that sends signals to the earth. Just as the orbiter and

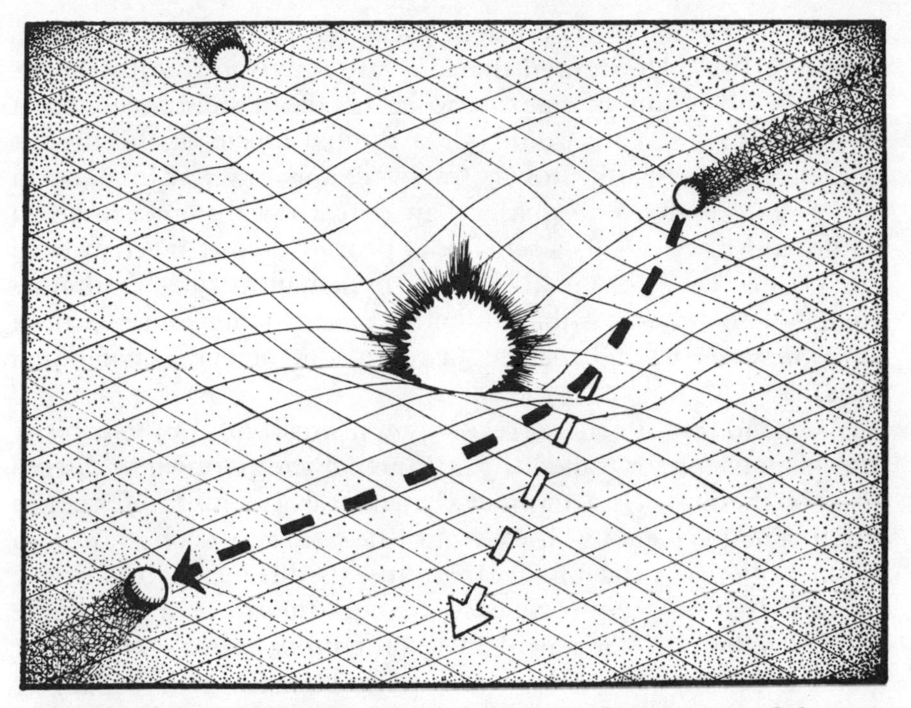

A schematic representation of the curvature of space around the sun. If a radar beam is reflected off a planet passing behind the sun as viewed from the earth, then the beam has to bend, causing a signal delay compared to what it would be if the beam did not pass near the sun. The different paths are indicated by the curved and straight dotted lines. The measured time delay between the curved and straight paths agrees with the general theory of relativity.

Mars are about to pass behind the sun from the viewpoint of the earth, the signals take longer and longer to reach earth because of the space curvature near the sun. Scientists can precisely measure the effective signal delays, and these, too, confirm Einstein's theory.

Perhaps the most dramatic confirmation of the bending of light was the discovery of a gravitational lens in 1979. Because mass causes space to curve in its vicinity, light paths bend near a large mass much as they do in an ordinary glass lens to achieve a focusing or distorting effect. Einstein predicted the gravitational lens effect in 1937. He showed that if a large mass acting like a lens existed on the line of sight between us and a still more distant light source, then we would see a double image of the distant source. Dennis Walsh, Robert Carswell, and Ray J. Weymann in 1979 noticed that one quasar—an extremely distant source of radio and light signals—actually appeared double when viewed by a powerful telescope. The best explanation for this double image of the quasar is that an entire galaxy, lying on the line of sight between us and the quasar, is producing the gravitational lens.

A second test of general relativity is a small shift in the orbit of the planet Mercury called the advance of the perihelion, discovered by the French astronomer Urbain Jean Joseph Le Verrier in 1859. The perihelion is the point of nearest approach to the sun in the elliptical orbit of a planet, and its advance refers to the amount that this point may move around the sun in a given time period. When Le Verrier calculated the influence of all the other planets on the advance of the perihelion of Mercury using Newton's law of gravity, he found about a 1 percent discrepancy between his theoretical calculations and the astronomical observations. Fortunately, he did not disregard this tiny discrepancy and published the result. Other scientists first attempted to discount this discrepancy from Newton's law, arguing that dust around the sun or the possibility that the sun was not perfectly round was responsible. But the dust was never seen and the sun was round. Einstein's general theory of relativity predicted small differences from Newton's law of gravity and gave a number of 43 seconds of arc per century—precisely the discrepancy Le Verrier had found! Today powerful radars can discriminate the mountains from the valleys on the surface of the planet Mercury. Such radars precisely measure Mercury's orbit, and again the perihelion shift agrees with general relativity.

The third test of general relativity is that clocks should run slower in a gravitational field. The stronger the force of gravity is, the slower time flows. Einstein, after all, said that time is what a clock measures. If a clock slows down, so does time. We actually age more slowly in a gravity field than someone in a gravity-free environment. This remarkable effect of the slowing of clocks is very small; only extremely precise clocks can detect it. The most precise clocks that have been made are atomic clocks, more accurate than the ancient standard of the movement of the stars. If two synchronized atomic clocks are left side by side, they will differ by only a fraction of a second after billions of years. We can check how gravity slows these clocks by putting one of the atomic clocks into an orbital trajectory high above the earth where gravity is weaker and then returning it to earth and comparing its elapsed time against a clock on earth where the gravity is relatively stronger. The observed time difference between the two clocks agrees with general relativity. In another version of this experiment, one atomic clock is taken from the National Bureau of Standards in Washington, D.C., near sea level, and moved to Denver, Colorado. The clock rates differ because of the difference in the gravitational force between the two locations, and these again accord with general relativity. By a tiny amount, people in Denver actually age more rapidly than those in Washington, D. C.

The three original tests of general relativity proposed by Einstein have been beautifully confirmed using modern technology. But beyond these predictions, the theory offers further implications which physicists are now investigating.

The theory of general relativity implies the existence of gravity waves, undulations of the curvature of space that propagate at the speed of light across any distance. It would be exciting to detect actual gravity waves, but most of the means of generating gravity waves from catastrophic cosmic events like stars exploding or colliding will generate gravity waves too weak to detect here on earth. One potential source of gravity waves could be black holes consuming stars at the core of our galaxy. Maybe, in a few decades, if there are strong enough gravity waves, we will detect them.

Recently the astrophysicists Hulse and Taylor's analysis of a binary pulsar has suggested some indirect evidence for gravity waves. A pulsar is a star collapsed to enormous density. A thimbleful of pulsar matter would weigh several tons. A binary pulsar is a pulsar orbiting around an ordinary star. Although we can't see the pulsar with an optical telescope, by using a large radio telescope we can detect the radio signals which the pulsar emits. In the case of a binary pulsar, the pulsar swings behind its companion star periodically, an event which blocks its radio signal. By measuring how often the signal is blocked, it's possible to determine the time or period of each orbit of the pulsar around its companion. Astronomers observing one such binary pulsar have measured its period over a number of years and observe that it is slowing down. How can we account for this slowing down?

The binary pulsar may be a gigantic transmitter of gravity waves. As it transmits gravity waves out into space, it loses energy, and this loss of energy is revealed by the pulsar's slowing down its orbital period. Using general relativity, astrophysicists have calculated the energy lost from gravitational waves radiating into space, and this is in remarkable agreement with the observed slowing down. Although indirect, the slowing down of the binary pulsar could be the first evidence for gravity waves.

These and other tests of general relativity have confirmed Einstein's theory. The tests reveal small but important differences from Newton's theory. That is because the gravity fields in our solar system are all weak, and for a weak field Einstein's and Newton's theories differ by only a small amount. Strong gravitational fields, such as those produced by totally collapsed matter in the form of black holes, reveal exciting new features of general relativity. With the strong gravity effects associated with black holes and the origin and expansion of the universe, general relativity finally comes into its own with qualitatively different features from Newton's theory. Through such discoveries we realize that it takes a revolution in our thinking to discover the new laws of nature. These new laws may at first yield only small corrections to the old ones. But the new laws have qualitative implications that extend far beyond the old ideas, as Einstein's theory of general relativity extends far beyond Newton's old theory. If we are ever to understand the beginning and end of the universe, we must go beyond Newton's theory of gravity into Einstein's general relativity.

General relativity, with its emphasis on geometry, opens up a new vision of the nature of the universe, providing the basis for cosmology, the study of the entire universe. For millennia, people have wondered about the universe and its origin. Now a new mathematical tool—the general theory of relativity—is available to cast these questions into a new form and perhaps even to answer them.

Looking out on a clear night we see the heavens filled with stars. The feeling is that we are very small; we know that the universe is far larger than even the visible stars suggest. All the stars we see are part of the Milky Way—our home galaxy—and this is just one of billions of galaxies. How can we study such vastness? We can imagine that the universe is like a gas in which the particles are galaxies. For this simplifying case of a uniform gas of galaxies we can solve the equations of general relativity.

The Soviet physicist Alexander Friedmann was the first to find these solutions to Einstein's equations. In 1922 he found the surprising result that Einstein's theory of general relativity implied the universe could not be static—it had to be changing. The gas of galaxies had to expand or contract. It would be as if our shadow friends had discovered that not only were they living in a curved space, but also that the curvature was changing in time.

Friedmann showed that if the density of the gas of galaxies was below a critical value, the universe was open and would continue to expand forever—the galaxies would move farther and farther apart. If the density of the galaxies was above a critical value, the universe would be closed and would eventually contract.

It is like throwing a stone. If you throw it up fast enough, above a critical velocity (related to the total matter in the earth), it will never return to earth—like the open universe, it never returns. Below that critical velocity the stone always returns to earth—like a closed universe. The best evidence that astronomers have today suggests that we are below the critical density for galactic matter and the universe is open. But should more matter be discovered, the actual density would increase and then we could have a closed universe which would expand and then contract.

At first Einstein didn't believe Friedmann's calculations and thought he'd made a mistake. Like most physicists and astronomers of this time, Einstein thought the universe was static and existed from an eternity in the past to an eternity in the future. A dynamic, evolving universe seemed contrary to experience and a gratuitous novelty. Because he wanted a closed, static universe, Einstein even went so far as to alter his relativity equations, adding a "cosmological term" that allowed for a static solution. He later called this mutilation "the biggest blunder of my life." So it was Friedmann, not Einstein, who had discovered that general relativity required an expanding, moving universe. His dramatic prediction was made seven years before the great cosmological discovery of the American astronomer Edwin Hubble. From a detailed study of distant galaxies, Hubble concluded that the universe was indeed expanding like a gigantic explosion. The universe was evolving!

The general theory of relativity was Einstein's greatest accomplishment; it represented the fulfillment of the classical, deterministic world view. While Einstein went beyond the physics of Newton, bringing the ideas of space, time, and matter to their modern form, the framework of his physics was completely deterministic. The great clockwork of Newton's universe was altered by Einstein—the gears and parts were different—but Einstein agreed with Newton that the motion of the clock was still completely determined into the infinite past and future.

It is difficult to imagine that a single person created general relativity. The theory combines the ideas of space, time, energy, matter, and geometry into a coherent whole of enormous scope and implication. How did Einstein invent general relativity?

While he was in Zurich and during his first years in Berlin, Einstein fell under the intellectual influence of the philosopher-physicist Ernst Mach, a major advocate of positivism in physics. Mach taught that theoretical physicists should never use any idea in physics which cannot be given a precise, direct meaning through experimental operations. Ideas without connection to the empirical world were deemed superfluous to physical theory. Mach's method became a guiding force in the development of the new physics. Einstein was a master of this method. Recall his definitions of space and time: Space is what we measure with a measuring rod; time is what we measure with a clock. These definitions, with their direct appeal to measurement, cut through all the excess philosophical baggage that the ideas of space and time had carried for centuries. The positivist insists that we talk only about what we can know through direct operations like a measurement. Physical reality is defined by actual empirical operations, not by fantasies in our heads.

However, after he settled in Berlin, Einstein moved away from the position of strict positivism, and this was only partly due to persuasive arguments offered by his colleague, Planck. It was as much Einstein's own success with the general theory of relativity and the method of thought he used to arrive at it that convinced him of the limitations of the strict positivist method. If Einstein had remained a positivist, I doubt that he would have discovered general relativity. Einstein subsequently described his own method in a letter to the philosopher Maurice Solovine, a friend from his days in Bern at the patent office. This method might be called "Einstein's postulational method."

In his letter to Solovine, Einstein included a diagram which illustrated his method. The diagram is:

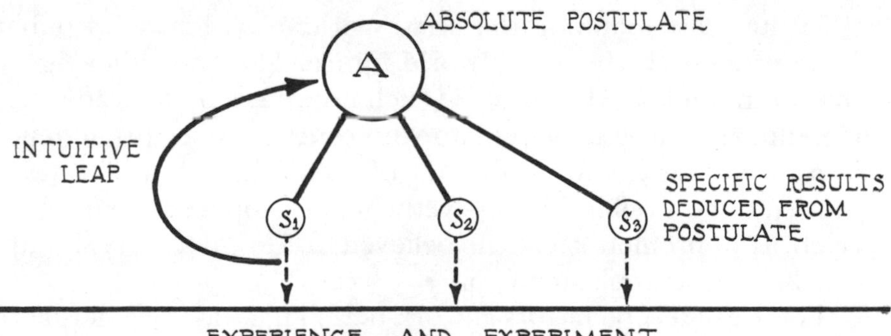

The scientist begins with the world of experience and experiments. On the basis of nothing more than physical intuition, he leaps from experience to abstracting an absolute postulate—just as Einstein realized that the equivalence principle implied gravity is geometry. Einstein made this conceptual leap far beyond where any experiment could check it and before he had any supporting evidence. How could there be such evidence? No physicist had even imagined the relation of gravity to geometry. The next step is to use the postulate to deduce specific theoretical results that can be experimentally checked. For general relativity, these results are the predictions such as the shift in the orbit of Mercury. Should an experiment falsify the theoretical results it also brings down the postulate on which they are based. This vulnerability of the absolute postulate to falsification is part of the positivist method.

But a strong antipositivist element central to Einstein's method is the intuitional leap from experience which sets up the absolute postulate in the first place. The theorist cannot rationally deduce the absolute postulate from experience, since it transcends experience. Only intuition, an inspired guess, can invent the postulate. This is what Einstein meant when he said, "For the creation of a theory, the mere collection of recorded phenomena never suffices—there must always be added a free invention of the human mind that attacks the heart of the matter." A great deal of creative work in physics proceeds by this method, which places intuition at the very first step, a nonrational but verifiable aspect of scientific creativity.

In the years following the First World War, Einstein's public fame rose and he became a world figure. The only other figure that I can think of who attracted such notice as a moral leader was Gandhi, and he was a statesman who courted public prominence as a means for leading India out of colonialism. Einstein never wanted to be a public figure—yet when it happened, he used his celebrity to promote causes he believed in. How does one account for this "Einstein phenomenon"?

There are several factors at work here. First was the emergence of the new media of radio and mass-circulation newspapers associated with the rise in literacy. Secondly, Europe was exhausted and devastated by the war; Germany especially had to salvage something from defeat. Popular attention turned to Einstein and his accomplishments, which seemed far removed from the political world and which reminded Germans of their great scientific culture. During the war, Einstein went his own way, apart as always. He was a pacifist at a time when that position was considered tantamount to treason. He was proud to be a Jew at a time when many German Jews were disguising their identity and assimilating. These were unpopular positions, but they established Einstein in the public eye as a man of principle at a time when men of principle were rare. Finally, that time in Europe was a period of ideological debate and conflict. In Russia, there was civil war, an aftermath of the Revolution of 1917. Everywhere fascism was on the rise. Social and religious philosophers sought support for their views in Einstein's new theories, which, it became clear, were the next step in the revelation of nature. Soviet physicists, led by V. Fock, found it necessary to defend relativity against charges of idealism and to point out that it was in strict accord with Lenin's materialism, the ideological basis for Soviet policy. Some scientists in England and America insisted that Einstein's relativity theory had nothing to do with moral or cultural relativism, a philosophy that maintained human moral values were relative to their social and cultural environment. This philosophy was then popular in the universities and was threatening to traditional religions. The astronomer Arthur Eddington, a Quaker himself, assured religious people there was still a place for God and Soul in the universe. In the face of these controversies, Einstein himself reiterated his cosmic philosophy already formulated when he was a teenager, that the universe was indifferent to mankind and its problems. But he asserted that moral questions were of utmost importance for human existence and that humanity must create a moral order for the sake of its survival.

Even as Einstein's eminence grew and his vision of the universe became part of public awareness, physics itself was moving ahead with enormous strides. In the 1920s, the quantum theory of atomic phenomena was created. Einstein rejected it not because it was wrong (it agreed with experiments) but because he felt it gave an incomplete description of physical reality and denied the objectivity and determinism of the world. His great debate with Niels Bohr began; but that is a story for another chapter. In the late 1920s and 1930s, a new generation of physicists emerged who accepted the new quantum theory and applied it with great success. The theory of the chemical bond was discovered; the new quantum theory explained the foundations of chemistry. The theories of solid-state matter, metals, electrical conductivity, and magnetization were developed out of the new quantum theory. Nuclear physics began.

Einstein had little to do with these developments. He was on the sidelines of physics after 1926. Einstein, in fact, thought that the new quantum theory was not radical enough. He thought the quantum theory might be a consequence of a unified field theory—a theory that combined electric, magnetic, and gravitational fields and went beyond general relativity. In 1938 he said, "I have now struggled with this basic problem of electricity for more than twenty years, and have become quite discouraged, though without being able to let go of it." Although he failed to unify electricity and gravity, he was one of the first physicists to emphasize the importance of seeking the unification of all forces in nature, a goal of physics on which great progress has occurred only recently. Of all his work, he felt that everything he had done would have been discovered without him except for general relativity. This was the crown of his creativity and of classical physics as a science. But the road to progress in physics, at least for the next half century, lay elsewhere.

My view is that after 1926, Einstein became involved in the mathematics of the unified field theory. For the rest of his life he could not resist the conceptual power and beauty of general relativity. The influence of this creation and the method of thought he used to arrive at it dominated all his subsequent thinking. He lost contact with "the old One" and the creative physical intuition he possessed for more than twenty years. The delicate balance between innocence and experience, prerequisite for creativity, tipped toward experience. As the physicist Paul Ehrenfest said when he heard of Einstein's opposition to the new quantum theory, "We have lost our leader."

Einstein held the classical view of determinism to the end of his life. For him, it was unthinkable that there was arbitrariness and chance in the fundamental structure of the universe. His vision of the cosmic code—the eternal laws of nature that govern all existence—left no room for chance or the intervention of human will and purpose. He felt that the quantum theory was superficial and that beyond the random play of atomic particles it described, we would find a new deterministic physics. While most other physicists are open to the possibility of revising the quantum theory, they do not believe that deterministic physics will ever return. Einstein, even as the leader of the physicists before 1926, was apart from them. After 1926, he was not only apart but also alone.

As a teenager I admired Einstein as a hero of science, a god in a distant pantheon of the intellect. Now through older eyes I see another side of him—his loneliness and emotional exile from capricious human feelings. He required that distance to forge the instruments of his immense genius, a genius which showed us a universe far greater and more bewildering than was previously imagined. His vision of the cosmos, stretching across the vast emptiness of space and time, indifferent to our humanity, now haunts us all. But I still wonder about the man that first saw this vision and look for clues to his character.

Einstein loved the music of Mozart. Both these men shared the sense of the ultimate vulnerability of all life but never lost their sense of play or a ready laugh. They knew that in this world the reality of life is that it need not be. Yet what we might learn from such men is that we too can celebrate our creative existence in the full awareness of its extinction. And that is the essence of irony.

With the rise of the tribal madness of National Socialism, Einstein left Germany on a forced emigration to the United States, a country he had already visited. Along with many other of the most brilliant European scientists, he brought to America a spirit of scientific inquiry centuries in the making. Talented Americans became willing students.

Einstein never felt at home in the United States. A product of the great German intellectual renaissance at the turn of the century, he did not adjust to the new ways. Once he remarked with his usual irony, "To the Jews I am a saint, to the Americans an exhibition piece, to my colleagues a mountebank."

Einstein knew that he had no choice in being born but his death could be his choice. Upon learning of his terminal illness, Einstein refused an operation. He died in Princeton, New Jersey, his American home, on April 18, 1955, attended by a nurse who could not understand his last words, which he spoke in German.

Michele Besso was Einstein's oldest friend from his days in the Bern patent office. They kept up a correspondence through fifty years. Einstein's unique acknowledgment in his 1905 paper on special relativity was for conversations with his friend Besso. What an honor! Einstein's friend died a month before he did, in Switzerland. Expressing his world view of absolute determinism, Einstein wrote movingly to Besso's son and sister, "Now he has departed from this strange world a little ahead of me. That means nothing. People like us, who believe in physics, know that the distinction between past, present, and future is only a stubbornly persistent illusion."

3. The First Quantum Physicists

By Zeus, Soddy, they'd have us out as alchemists!
—Ernest Rutherford

As a college freshman I had my first contact with the quantum theory by purchasing a copy of *Quantum Mechanics* by Leonard Schiff, who became my teacher in graduate school. I read his book and worked out the problems. Quantum mechanics was for me an exercise in solving differential equations. To my freshman's mind, unencumbered by any bias from the older, classical physics, the quantum theory presented no problems. It was simply an abstract mathematical description of atomic processes. I had no sense of the "quantum weirdness" of the atomic world; it was the earlier theory of special relativity, with its space contractions and time dilations, that seemed bizarre to me. But as I continued my study my reaction reversed—relativity seemed less bizarre and more in accord with common sense, while the quantum theory seemed more and more "weird." Pursuing the mathematics of the quantum theory, I felt pushed beyond common sense into unimagined areas. Later, I found out that my experience paralleled that of the physicists who first discovered the new quantum theory. First they discovered the mathematical equations of the quantum theory which worked experimentally; then they pondered the equations and their meaning for the real world, developing an interpretation which departed radically from naive realism. As I realized what the abstract mathematics of quantum theory was actually saying, the world became a very strange place indeed. I became uncomfortable. I would like to share that discomfort with you.

What is this quantum weirdness? The physics of the new quantum theory can be contrasted with the older Newtonian physics which it replaced. Newton's laws brought order to the visible world of ordinary objects and events like stones falling, the motion of the planets, the flow of rivers and the tides. The primary characteristics of the Newtonian world view were its determinism—the clockwork universe determined from the beginning to the end of time—and its objectivity—the assumption that stones and planets objectively exist even if we do not directly observe them; turn your back on them and they are still there.

In the quantum theory these common-sense interpretations of the world (like determinism and objectivity) cannot be maintained. Although the quantum world is rationally comprehensible, it cannot be visualized like the Newtonian world. And that is not just because the atomic and subatomic world of quanta is very small, but because the visual conventions we adopt from the world of ordinary objects do not apply to quantum objects. For example, we can visualize that a stone can be both at rest and at a precise place. But it is meaningless to speak of a quantum particle such as an electron resting at a point in space. Furthermore, electrons can materialize in places where Newton's laws say they can't be. Physicists and mathematicians have shown that thinking about quantum particles as ordinary objects is in conflict with experiment.

Not only does quantum theory deny the standard idea of objectivity but it also has destroyed the deterministic world view. According to the quantum theory, some events such as electrons jumping around atoms occur at random. There just isn't any physical law that will ever tell us when an electron is going to jump; the best we can do is to give the probability of a jump. The smallest wheels of the great clockwork, the atoms, do not obey deterministic laws.

The inventors of the quantum theory found yet another contrast with the Newtonian world view—the observer-created reality. They found that the quantum theory requires that what an observer decides to measure influences the measurement. What is actually going on in the quantum world depends on how we decide to observe it. The world just isn't "there" independent of our observing it; what is "there" depends in part on what we choose to see—reality is partially created by the observer.

These properties of the quantum world—its lack of objectivity, its indeterminacy, and the observer-created reality—which distinguish it from the ordinary world perceived by our senses I refer to as "quantum weirdness." Einstein resisted quantum weirdness, especially the notion of an observer-created reality. The fact that an observer was directly involved with the outcome of measurements clashed with his deterministic world view that nature was indifferent to human choices.

Something inside of us doesn't want to understand quantum reality. Intellectually we accept it because it is mathematically consistent and agrees brilliantly with experiment. Yet the mind is not able to rest. The way in which physicists and others have trouble grasping quantum reality reminds me of the way children respond when confronting a concept they do not yet grasp. Jean Piaget, the psychologist, studied this phenomenon in children. If a child of a certain age is shown a collection of transparent vessels all of very different shape filled with a liquid to the same level, the child thinks that all the vessels have the same amount of liquid. The child does not yet grasp that the amount of liquid has to do with volume, not just height. If the correct way of viewing the problem is explained to the child, the child will often understand it but immediately reverts back to the old way of thinking. Only after a specific age, around six or seven years, is the child able to grasp the relation between amount and volume. Coming to grips with quantum reality is like that. After you think you have grasped it and some picture of quantum reality forms in your mind, you immediately revert back to the old, classical way of thinking, just like the children in Piaget's experiment.

It is important to realize that the microworld of atoms, electrons, and elementary particles is not entirely unlike the classical world, the physical world of naive realism. A single atom can be isolated in a box; electrons and other particles leave tracks on photographic emulsions or in cloud chambers. We can push them around using electric and magnetic fields. Experimentalists can measure certain properties of these tiny objects, such as their mass, electric charge, spin, and magnetization. Physicists, like most people, think of microworld particles in just that way. They are just tiny little things. We can make particle beams of them or bounce them off each other and make them dance to our tune. Where is the quantum weirdness? What is so hard to grasp?

The quantum weirdness comes when you start to ask certain kinds of questions about atoms, electrons, and photons. And it comes only when you ask these special questions and set up experiments to try to answer them. For example, if you try to measure precisely both the position of an electron and its velocity by repeated measurements, you find it can't be done. Every time you measure its position the velocity changes, and vice versa—the electron has a kind of quantum slipperiness. If the electron were an ordinary object, you would be able to determine simultaneously both its position and its velocity. But the electron is a quantum particle, and the ordinary idea of objectivity fails. Until you start asking detailed questions about quantum particles, such as what is the precise position and velocity of a particle, you can live happily in a paradise of naive realism.

Once a person recognizes that the quantum weirdness of the microworld is unavoidable, he can take two attitudes. The first is to forget it and stick to the mathematics of quantum theory. In that way he will find the right answers and make progress in discovering the laws of the microworld. Most theoretical physicists, following the lead of Paul Dirac and Werner Heisenberg, who laid the mathematical foundations of the new quantum theory, take this attitude. The second attitude is the philosopher's approach, which tries to interpret the quantum weirdness of the microworld in terms of physical reality. They are interested in developing a conceptual picture of the quantum world that is intelligible as well as mathematically consistent. Niels Bohr founded that attitude for modern physics, and he had much to say about the interpretation of reality.

The story of the discovery of the quantum theory began with Max Planck's determination of the black-body radiation law, the giant first step in 1900. The main feature of the old quantum theory was that it represented attempts on the part of physicists to fit the idea of Planck's quantum—that there was a discrete element in nature—into classical Newtonian physics. In his work on black-body radiation, Max Planck introduced a new constant into physics, called h, which was a measure of the amount of discreteness in atomic processes. When Planck did his work, in 1900, physicists thought that atoms could have any value for their total energy—energy was a continuous variable. But Planck's quantum hypothesis implies that energy exchange was quantized. Although the introduction of a quantum of energy had no basis in classical physics, it was not yet clear that the new theory required a radical break with classical concepts. Theoretical physicists first tried to reconcile Planck's quantum hypothesis with classical physics.

Physicists are conservative revolutionaries. They do not give up tried and tested principles until experimental evidence—or an appeal to logical and conceptual simplicity—forces them into a new, sometimes revolutionary, viewpoint. Such conservatism is at the core of the critical structure of inquiry. Pseudoscientists lack that commitment to existing principles, preferring instead to introduce all sorts of ideas from the outside. Werner Heisenberg commented, "Modern theory did not arise from revolutionary ideas which have been, so to speak, introduced into the exact sciences from without. On the contrary, they have forced their way into research which was attempting consistently to carry out the program of classical physics—they arise out of its very nature." The old quantum theory represented a program to reconcile the quanta with classical physics.

Einstein took up Planck's idea in his 1905 paper on the photoelectric effect. Planck assumed the sources of light exchanged quantized energy. Einstein, going a step further, assumed that light was itself quantized—light consisted of particles called photons. This revolutionary idea broke with the then well-established wave theory of light—reason enough for most physicists to reject it. Other physicists resisted Einstein's proposal because it only explained the photoelectric effect, which was hardly direct evidence for the photon. But Einstein held firm to the notion of a wave-particle duality for light and attempted reconciling these apparently contradictory properties of light but without success.

The theoretical ideas of Planck and Einstein which advanced the quantum theory were a response to experiments which opened a whole new realm of natural phenomena. By the end of the nineteenth century a great number of puzzling new properties of matter were discovered; for the first time, scientists were making direct contact with atomic processes. Roentgen discovered the penetrating X-rays in 1895. Henri Becquerel discovered radioactivity in 1896, and the Curies isolated radium in 1898. In 1897, J. J. Thomson discovered the electron, a new elementary particle. A puzzling discovery was that under certain circumstances atoms emit spectral lines of light. If a substance is heated or if an electric current is passed through a gas of atoms, the substance or gas will emit light. If the spectrum of the light is analyzed by a prism that splits off the various colors, only definite colored lines appear in the spectrum. Neon colored lights offer one example. Each chemical element has a definite and unique set of colored lines, called its line spectrum. No one had any explanation for this phenomenon in the nineteenth century. Yet here lay the experimental clue to the structure of the atom.

Ernest Rutherford was already a famous experimentalist for his discovery of the radioactive transformation of elements with Frederick Soddy when he came to Manchester University. Rutherford and Soddy had found that chemical elements, previously thought to be immutable, changed in the process of radioactivity. Soddy suggested they should call the new process "radioactive transmutation." Transmutation of the elements, such as lead into gold, was an ancient alchemical dream already discredited by nineteenth-century chemists and physicists. Soddy's suggestion was met by Rutherford's sharp reply, "By Zeus, Soddy, they'd have us out as alchemists!" But in fact they had discovered transmutation of the elements.

At Manchester, Rutherford was studying alpha particles, stable positively charged helium nuclei that are emitted by radioactive substances. Rutherford, who did not have the patience to do long hours of counting scintillations on a screen that detected alpha-particle bombardment, unleashed a young assistant, Marsden, on an experiment. The experiment is beautiful for its simplicity. A radioactive source of alpha particles is placed near a metal foil (Marsden used gold foil). The alpha particles are projectiles, like little bullets being fired at the foil. Most of the alpha particles go straight through the foil and are detected on a screen. However, on a hunch, Rutherford asked Marsden to look for alpha particles that were strongly scattered by the foil and widely deflected. By placing the detecting screen away from the line of sight to the alpha source, Marsden found a few deflected alphas. He observed that some even scattered back toward the alpha source. It would be like firing bullets at a piece of tissue paper only to find some bullets bounced backward. This discovery initiated a series of experiments.

What caused some alpha particles to scatter backward from the gold foil? Rutherford knew the alpha particles were positively charged. In the gold foil these particles would sometimes pass close to the atomic nuclei, also positively charged. Since like charges repel each other, this caused the large deflections of some alpha particles off the atomic nuclei. By carefully studying these deflections, Rutherford determined the major features of atomic structure. A window on the microworld opened.

The idea of atoms, held by many people, was that they were without parts, completely elementary, the end of all material structure—a building block for the rest of matter. While a few theoretical physicists speculated about the possibility of atomic structure, there was no experimental support for such speculations. Rutherford's simple scattering experiment gave humankind its first glimpse into the structure of the atom.

The picture of the atom Rutherford announced in May 1911 was that most of the mass of the atom was concentrated in a tiny, positively charged core, later called the nucleus, while the negatively charged electrons, with very small mass, formed a large cloud about the nucleus, accounting for the size of the atom. The massive nucleus was ten thousand times smaller than the atom. Rutherford's atom was like a little solar system with the nucleus as the sun and the electrons as the planets and with electric forces instead of gravity binding the system together.

Although Rutherford's scattering experiments were compelling, from the standpoint of classical physics his planetary picture of the atom was completely unstable. According to classical physics, the electron in orbit about the nucleus should radiate away its energy in the form of electromagnetic waves and fall rapidly into the nucleus. Physicists knew that according to the laws of classical physics, Rutherford's atom ought to collapse. But there it was nevertheless. This unsatisfactory situation soon changed dramatically. Around 1912, Rutherford wrote from Manchester to his friend Boltwood, "Bohr, a Dane, has pulled out of Cambridge and turned up here to get some experience in radioactive work." Niels Bohr, a student of J. J. Thomson's at Cambridge, actually spent less than half a year at Manchester before returning to his native Copenhagen. However, in spite of his brief visit, Rutherford made an impact on the young Dane.

Bohr, challenged by the problem of atomic structure, took an imaginatively daring step: He simply dispensed with some of the rules of classical physics and instead applied the quantum theory of Planck and Einstein to the problem of atomic structure. Remarkably, the few features of the quantum theory already known at the time could solve the problem—as long as one did not worry about the conflict with classical physics. Bohr simply assumed that the electrons in orbit about the nucleus do not radiate light and that the light emitted by atoms is due to some other physics. He showed that Planck's idea of energy quantization implied that only specific orbits for the electrons are allowed. In order to ensure the stability of atoms, Bohr postulated a lowest orbit beyond which the orbiting electron could not fall. When an electron drops from a higher orbit to a lower one, thereby losing energy, the atom containing that electron emits light, which carries off the lost energy. Because only certain electron orbits are allowed, only certain jumps of the electron between orbits can take place, and consequently the energy of the emitted light is quantized. Since the energy of light is related to its color, only specific colors of light can be emitted by atoms. In this way Bohr's theoretical model of the atom accounts for the existence of the mysterious spectral lines. The experimentally observed fact that each different atom emitted light with unique and distinct colors revealed the quantum structure of atoms.

One way of imagining the energy levels of Bohr's atom is to think of a musical stringed instrument like a harp. Each string when plucked has a definite vibration or sound. Similarly, when an electron jumps orbits in an atom there results the emission of a light wave with a definite vibration or color. That is the origin of the discrete light spectrum.

Bohr applied his novel ideas to the simplest atom, hydrogen, which consists of a single proton with a single electron in orbit about it. The advantage of studying such a simple atom is that the allowed orbits of the electron could be precisely calculated and hence the spectrum of light from hydrogen determined. Bohr's calculations of the hydrogen light spectrum based on his theoretical model of the atom agreed adequately with the experimentally observed spectrum. Such agreement between theory and experiment could not be an accident. It meant that the combination of ideas Bohr took from the quantum theory really worked—the scientific imagination made its first successful step into the quantum structure of the atoms. The ancient capability of the human mind to comprehend a new environment, in this case the atomic structure of matter, was again powerfully reinforced.

Theoretical physicists seized Bohr's ideas and applied them to more complicated atoms. But Bohr's model, like every great scientific advance, raised many new questions—questions that couldn't be asked before. When does an electron change its orbit and cause light to be emitted from the atom? What causes a particular jump? What direction does the emitted light take off in, and why? These questions troubled Einstein. According to classical physics, the laws of motion precisely determine the future behavior of a physical system like an atom. But atoms emitting light seemed to behave spontaneously and undeterminedly. Atoms jump. But why and in what direction? The same spontaneity, Einstein realized, characterized radioactivity.

At first, physicists tried to fit the behavior of atoms into the framework of the classical theory of electromagnetism and made desperate attempts to answer the enigma of the quantum jumps without using light quanta. In 1924, Niels Bohr, Hendrick Kramers, and John Slater wrote an article advocating this approach at the expense of abandoning the laws of energy and momentum conservation at the level of the atom—a revolutionary proposal, because these laws are among the most well-tested physical laws. At the time of this proposal there had been no direct experimental evidence that these conservation laws worked for individual atomic processes. However, it soon came. Arthur H. Compton and A. W. Simon scattered individual photons, the light particles, from electrons. Using a Wilson cloud chamber, a device that displayed the tracks of individual electrons, they verified to a high degree of accuracy the conservation laws for individual atomic processes. For most physicists, these experiments done in 1925 vindicated Einstein's 1905 proposal of the light quantum.

Through a multitude of new atomic experiments such as Rutherford's and Compton's, the structure of the atom was revealed. These experiments forced theoretical physicists into a new and unfamiliar world; the usual rules of classical physics no longer seemed to work. In the atom the human mind was shown a new message—a new physics revealed in the structure of the atomic microworld. The world view of determinism, supported by centuries of experiment and physical theory, was about to fall.

Bohr accepted the results of the experiments of Compton and of Simon, both the correctness of the conservation laws and the existence of the light quantum or photon. He concluded, in July 1925, "One must be prepared for the fact that the required generalization of the classical electrodynamical theory demands a profound revolution in the concepts on which the description of nature has until now been founded." Bohr was ready for the revolution. It soon came. The first shot had already been fired on a small island in the North Sea.

4. Heisenberg on Helgoland

If God has made the world a perfect mechanism, He has at least conceded so much to our imperfect intellect that in order to predict little parts of it, we need not solve innumerable differential equations, but can use dice with fair success.

—Max Born

Helgoland is a small island in the North Sea, not far from the North German industrial city of Hamburg, with high red cliffs and fresh sea breezes. It was here that Werner Heisenberg invented matrix mechanics—the first step of the new quantum theory. Heisenberg belonged to a new generation of physicists who came out of the First World War with a different outlook, which included a distrust of the previous generation. He was among the many German students who set out to find something of value, something uncorrupted by the recent past. His father, a classicist, instilled in him a love of Greek philosophy and literature. The young Heisenberg with his clear eyes, brush-cut hair, shorts, and keen sense of competition was the image of the postwar German youth movement. In spite of his strong attraction to classics, Heisenberg was drawn to science. He went to study in Munich with Arnold Sommerfeld, who invited him in 1921 to hear Niels Bohr lecture in Göttingen at what was known as the "Bohr Festival." Heisenberg was tempted to become a pure mathematician; but through long discussions with Bohr he became fascinated with the problem of atomic theory and decided instead to become a theoretical physicist. Heisenberg realized that the realm of abstract mathematics could be applied to the most difficult new problems in physics—a connection between pure ideas and the real world that excited him. Reflecting on this, Heisenberg later said, "I also learned something perhaps even more important, namely that in science a decision can always be reached as to what is right and wrong. It is not a question of belief, or Weltanschauung, or hypothesis; but a certain statement could be simply right and another statement wrong. Neither origin nor race decides this question: It is decided by nature, or if you prefer by God, in any case not by man." Like Einstein a generation before him, Werner Heisenberg had encountered the cosmic code, the internal logic of the universe. Through physics he could know the very soul of the universe, a knowledge far removed from those political events which had caused so much recent human suffering. After completing his doctoral work with Sommerfeld in 1924, Heisenberg went to join Bohr in Copenhagen and work on the new atomic theory.

Bohr had always wanted a place, like Rutherford's lab in Manchester which he had visited, where physicists could discuss their problems without the formal student-professor relation interfering. In 1920, through donations from Danish businesses, including the Carlsberg brewery, Bohr realized his vision and founded an institute in Copenhagen which became known as the Niels Bohr Institute. Bohr gathered about him young, bright students from Europe, America, and the Soviet Union to study the problems of atoms. Here Heisenberg found an intellectual environment that challenged his creative power—a community of geniuses that would soon become the new scientific establishment. These students were a brilliant, arrogant, and penniless lot. The general public had little interest in or understanding of their work, but this lack of attention did not discourage them. They were convinced they were creating a scientific revolution which would transform the understanding of reality.

After a year with Bohr, Heisenberg left to become an assistant to Max Born, director of the physics institute at Göttingen University in Germany. Like many physicists, Heisenberg was wrestling with the puzzle of atomic spectral lines. He was also struggling with a bout of spring hay fever in Göttingen and decided to go to Helgoland to clear his head. Here the lightning struck, and in one day and night Heisenberg invented a new mechanics. His paper was completed in July 1925. Similar to Planck's earlier idea of the quantum in 1900, there was no historical precedent for Heisenberg's idea. A single rock had been loosened by the lightning. An avalanche followed.

Heisenberg was interested in Greek philosophy, especially Plato and the atomists, who thought of atoms conceptually, not as things with parts. Most physicists tried to make physical pictures of atoms, but Heisenberg, like the Greeks, felt it was necessary to dispense with all pictures of atoms, of electrons circulating about the nucleus with definite radii like little solar systems. He did not think about what atoms were but what they did—their energy transitions. Proceeding mathematically, he described the energy transitions of an atom as an array of numbers. Applying his remarkable mathematical resourcefulness, he found rules that these arrays of numbers obeyed and used these rules to calculate atomic processes. Before leaving again for Copenhagen, he showed his work to Max Born.

Born recognized in Heisenberg's arrays of numbers the mathematics of matrices. A matrix is a generalization of the idea of a simple number to a square or rectangular array of numbers. Consistent algebraic rules for multiplying and dividing such matrices had been worked out by the mathematicians. Born solicited the aid of his student Pascual Jordan, and together they worked out the details. Born and Jordan wrote a paper extending Heisenberg's ideas, pointing out the importance of matrix algebra for describing atomic energy transitions. Somehow matrices rather than simple numbers were the correct language for describing the atom.

In classical physics the physical variables that describe the motion of a particle are simple numbers. For example, the position (q) of a particle from a fixed point might be 5 feet (q = 5); its momentum (p = mass of the particle times its velocity) might be designated by 3 ($p = 3$). Simple numbers like 5 and 3 obey the commutative law of multiplication: that is, $3 \times 5 = 5 \times 3 = 15$ —the order of multiplication does not matter. Likewise for the position and momentum of a particle in classical physics; these variables, since they were always simple numbers, obeyed the commutative law, $p \times q = q \times p$.

The main idea of the new matrix mechanics is that physical variables like the position q and momentum p of a particle were no longer simple numbers but *matrices*. Matrices do not necessarily obey the commutative law of multiplication—$p \times q$ does not have to equal $q \times p$. Born and Jordan's paper contained a relation for the matrices that represented the position q and momentum p of a particle which implied that the difference between $p \times q$ and $q \times p$ was proportional to Planck's constant, h. If we lived in a continuous world in which h was zero, then the matrices p and q would obey the commutative law like simple numbers—just as in the old classical physics. But because h was different from zero, although very small in the real world, the position q and momentum p of a particle could no longer be thought of as simple numbers—they had to be represented as matrices and obeyed the noncommutative laws of the new matrix mechanics, not the commutative law of classical mechanics. What could this possibly mean? Physicists, like most people, think of the position of a particle as having a definite value specified by a simple number. But in the new matrix mechanics the position of a particle was described by a matrix—not a simple number. What then was the "real" position of a quantum particle? Here, for the first time, arose the astonishing problem of physically interpreting the mathematics of the new mechanics, a problem quantum physicists would be struggling with in the coming years.

Heisenberg, in Copenhagen, when he heard of the recent work of Born and Jordan, did not know what a matrix was; but he quickly learned. Later in the same year, 1925, Heisenberg visited the Cavendish Laboratory at Cambridge, England, and gave a seminar on his recent work in the study of Peter Kapitza, a visiting Soviet experimentalist. In the audience was Paul Dirac, twenty-three years old and a brilliant mathematical physicist. Dirac understood the essence of Heisenberg's work immediately. Soon after Heisenberg left Cambridge, Dirac wrote a lucid paper formulating the new matrix mechanics and showed how it was a complete dynamical theory that replaced classical mechanics.

Meanwhile, in Göttingen, Born and Jordan, working in collaboration by letter with Heisenberg in Copenhagen, arrived at the same conclusions but by a slightly different route. The two papers, one by Dirac and the other co-authored by Born, Jordan, and Heisenberg—both sparked by Heisenberg's insight on Helgoland—mark the beginning of matrix quantum mechanics.

The new matrix mechanics was the mathematical modification of Newton's classical mechanics that physicists sought—it provided a mathematical description of moving particles, just like the earlier classical theory. But it also went beyond it. Theoretical physicists had created a new mathematical theory, and now, with great excitement, they turned to face the question: Did the new theory actually describe nature—was matrix mechanics the right quantum theory of the atom?

Heisenberg, in Copenhagen, worked hard to apply the new matrix methods to determining the light spectrum of the hydrogen atom. Bohr had already solved this problem, but it was of interest to see if the new method would yield the same result. The solution of this problem fell to the brash and brilliant young physicist Wolfgang Pauli. One colleague remarked of Pauli that it was not possible to distinguish his rudeness from his politeness. He was a ruthless critic of ideas, sometimes signing his letters "The Wrath of God." While a student with Arnold Sommerfeld in Munich, Pauli had already acquired a scientific reputation for a clear encyclopedia article he had written on the special theory of relativity. Once Einstein came to lecture at Munich and at the end of the lecture the nineteen-year-old Pauli stood up and said, "You know, what Mr. Einstein said is not so stupid. . . ." Later when he went to Copenhagen to work with Bohr, Pauli would get into long discussions with Bohr. Once at the end of a heated argument with Bohr he said to Bohr, "Shut up! You are being an idiot." "But

Pauli . . ." Bohr protested. "No, it's stupidity. I will not listen to another word." That is the kind of man he was. No intellectual dissembler or shoddy thinker could last long with Pauli around; unfortunately even physicists with correct ideas could be defeated by Pauli if he thought they were wrong.

Pauli quickly mastered the mathematics of matrices, solved the problem of the light spectrum of the hydrogen atom, and obtained the same result that Bohr had found earlier. Pauli also determined the light spectrum of a hydrogen atom placed in an electric or magnetic field, a problem that had previously resisted solution. The power of the new matrix mechanics was evident.

Physicists didn't get a picture of the atom or of quantum processes from the new matrix mechanics—it was invented precisely to avoid making a physical picture. The attitude of Dirac and Heisenberg was that a consistent mathematical description of nature was the road to truth in physics. The need to visualize the atomic world was a holdover from classical physics not appropriate to the new matrix theory. Many physicists felt dissatisfied with this attitude, and while Bohr, Born, Jordan, Heisenberg, Dirac, and Pauli were working on the new matrix mechanics, an alternate theory of the atom was developed, resulting in the invention of wave mechanics.

Einstein, we recall, theorized that light was a particle in 1905, an idea which flew in the face of the fact that light was an electromagnetic wave. As early as 1909, he suggested that the future theory of light would fuse the particle and wave theory of light, but there had been little progress in this direction. It seemed that light should be either a particle or a wave.

The next step was taken by Louis de Broglie, a French prince whose intellectual interests took him to the frontiers of physics. He reasoned by analogy that if light, which seemed so clearly to be a wave, could sometimes behave like a particle—the photon—then an electron, clearly a particle, could sometimes behave like a wave. The crucial ideas were presented in two papers of September 1923, in which de Broglie deduced the wavelength of the electron. He suggested that his idea could be experimentally confirmed if electrons showed diffraction phenomena like true waves. Diffraction of a wave around an obstacle, like an ocean wave hitting the edge of a jetty, refers to its bending behind an obstacle, unlike a particle beam, which casts sharp shadows. Sound is a wave, and that is why we can hear around corners; it "bends" around them. The papers became de Broglie's doctoral thesis, and a copy of the thesis was sent to Einstein by an examiner, Paul Langevin. Einstein thought highly of the thesis and did much to bring de Broglie's novel ideas to the attention of other physicists.

One of the physicists who heard about de Broglie's electron waves was the Austrian Erwin Schrödinger. He pondered the significance of the wave idea and devised an equation that the electron wave shape would have to obey if the electron was part of the hydrogen atom. Using his equation, he deduced the light spectrum of hydrogen—it was the same that Bohr had found years earlier. The strange notion that the electron was a wave was quantitatively vindicated. Schrödinger's paper appeared in January 1926 and marks the beginning of wave mechanics, another completely general way of formulating the new mechanics of the atom.

The "Schrödinger equation" applied to all sorts of quantum problems. A series of experiments supported Schrödinger's and de Broglie's thesis that electrons exhibited diffraction—there was no doubt that true waves were involved. But waves of what? The problem of the interpretation of the de Broglie—Schrödinger waves became the central puzzle of the new wave mechanics.

Schrödinger himself offered one of the first interpretations: The electron is not a particle, he argued, it is a matter wave as an ocean wave is a water wave. According to this interpretation the particle idea is wrong or only approximate. All quantum objects, not just electrons, are little waves—and all of nature is a great wave phenomenon.

This matter-wave interpretation was rejected by the Göttingen group led by Max Born. They knew that one could count individual particles with a Geiger counter or see their tracks in a Wilson cloud chamber. The corpuscular nature of the electron—the fact it behaved like a true particle—was not a convention. But what, then, were the waves?

It was Max Born himself who answered that perplexing and crucial question. His interpretation marks the birth of the God who plays dice and the end of determinism in physics. It occurred in June 1926, six months after Schrödinger's paper, and profoundly distressed the community of physicists. Born interpreted the de Broglie–Schrödinger wave function as specifying the probability of finding an electron at some point in space.

Think of a wave moving through space. Sometimes the wave height is just higher than the average level and sometimes lower. The height of the wave is called its amplitude. What Born said was that the square of the wave amplitude at any point in space gave the probability for finding an individual electron there. For example, in regions of space where the wave amplitude is large, the probability of finding an electron there is also large; perhaps one out of two times an electron will be found there. Similarly, where the wave amplitude is small the probability of finding the electron is also small—say one in ten. The electron is always a true particle and its Schrödinger wave function only specifies the probability for finding it at some point in space. Born realized the waves are not *material,* as Schrödinger wrongly supposed; they are waves of *probability,* rather like actuarial statistics for the creation of individual particles that can change from point to point in space and time. This description of the motion of quantum particles is inherently statistical—it is impossible to track them precisely. The best physicists could do was to establish the probable motion of a particle. Born demonstrated the consistency of his interpretation by a careful analysis of atomic collision experiments.

How are we supposed to think about the atomic world of the quanta? Atoms, photons, and electrons really exist as particles, but their properties—such as their location in space, momentum, and energy—exist only on a contingency basis. Imagine that an individual atom is a deck of cards and a specific energy level of

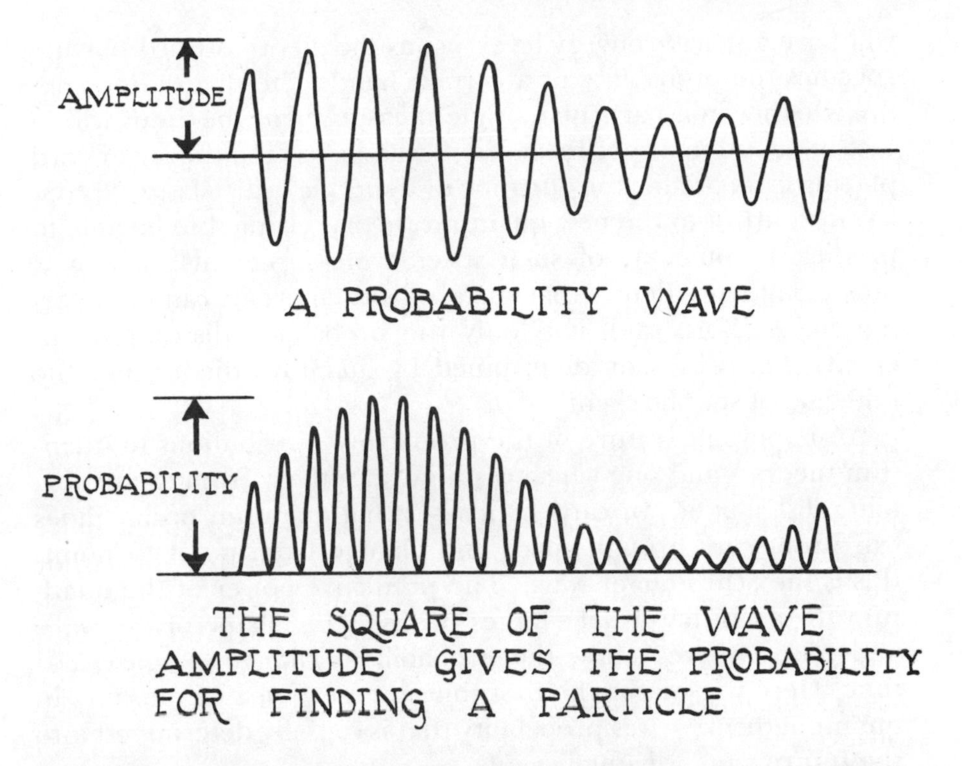

According to Max Born's statistical interpretation of the de Broglie-Schrodinger wave, the height or amplitude of the wave when squared gives the probability for finding the particle at that position. All that quantum theory could do was predict the wave shape and hence the probability that a quantum particle would have certain properties; it could not predict with certainty the outcome of single measurements of those properties, as did the old classical physics.

that atom corresponds to a specific poker hand dealt from the deck. Poker hands have probabilities that can be calculated—using the theory of card playing it is possible to determine precisely the possibility of a given hand's being obtained from the dealer. The theory does not predict the outcome of a particular deal. Demanding this latter kind of determinism requires looking into the deck—cheating. According to Born, the de Broglie–Schrödinger wave function specifies the probability that an atom will have a specific energy level just as the theory of card playing specifies the probability of a certain hand. The theory does not say whether in a particular single measurement the atom will in fact be found in a specific energy level, just as the theory of card playing can't predict the outcome of a specific deal. Classical physics, in contrast to the new quantum theory, claimed to be able to predict the outcome of such specific measurements. The new quantum theory denies that such individual events can be determined. As Born said, it is only the probability distribution of events that is causally determined by quantum theory, not the outcome of specific events.

An important feature of the probability distributions in quantum theory—and one which distinguishes them from the probability distributions of card hands—is that quantum probabilities can propagate through space and change from point to point; this is the Schrödinger wave. The predictive power of the quantum theory is that it determines precisely the shape of the wave and how it moves—how the probabilities change in space and time. Here we see for the first time the new idea of causality in quantum theory—it is probability that is causally determined into the future, not individual events.

Born was excited by his statistical interpretation of the wave theory but found himself alone. When Schrödinger heard of Born's interpretation, he remarked that he might not have written his paper if he had known the consequences—he never accepted indeterminism. Max Planck agreed with Schrödinger's matter waves and when Schrödinger accepted Planck's chair in Berlin, the retiring Planck praised him as the man who had brought determinism back to physics. In late 1926, Einstein wrote to Born, "Quantum mechanics is certainly imposing. But an inner voice tells me it is not yet the real thing. The theory says a lot, but does not really bring us closer to the secret of 'the old One.'" Born was especially disappointed by Einstein's rejection of the statistical interpretation. But Born was right.

This indeterminism was the first example of quantum weirdness. It implied the existence of physical events that were forever unknowable and unpredictable. Not only must human experimenters give up ever knowing when a particular atom is going to radiate or a particular nucleus undergo radioactive decay, but these events are even unknown in the perfect mind of God. Physicists, irrespective of their belief, may invoke God when they feel issues of principle are at stake because the God of the physicists is cosmic order. The indeterminism of the quantum theory is an issue of principle, of what is knowable and unknowable, not experimental technique—that is what distressed Einstein. Even God can give you only the odds for some events to occur, not certainty. About this time Einstein started stating his objection to the new quantum theory with the remark that he didn't believe God plays dice. Max Born, who always considered Einstein his physics master, later responded, "If God has made the world a perfect mechanism, He has at least conceded so much to our imperfect intellect that in order to predict little parts of it, we need not solve innumerable differential equations, but can use dice with fair success." The door to the indeterminate universe opened.

So there were now two explanations of atomic phenomena, Heisenberg's matrix mechanics and Schrödinger's wave mechanics. How could this be? It was Dirac who showed that the matrix and wave mechanics were completely equivalent through his transformation theory—they were simply different representations within the same theory. Physicists refer to them as Heisenberg (matrix) representation and the Schrödinger (wave) representation.

A good way to get across the significance of Dirac's transformation theory is by the analogy between language and mathematics. They are both symbolic means of representing the world; language is richer, while mathematics is more precise. Suppose someone describes a tree in the English language while someone else describes it in Arabic. The English and Arabic descriptions are different symbolic representations of the same object. If you want to describe the tree, you must pick at least one language or representation. Once you have one representation you can find the others by the rules of translation or transformation. That is how it is in the mathematical description of quantum objects like electrons. Some representations emphasize the wavelike properties, others the particlelike properties, but it is always the same entity that is being represented. That different representations are subject to laws of transformation is a profound idea. It is by varying the symbolic representations through transformations that we arrive at the notion of *invariants*: those deep, intrinsic properties of an object which are not just artifacts of how we describe it. We learn what makes a tree in any language. Invariants establish the true structure of an object.

Wave mechanics and matrix mechanics employ different representations to describe the same behavior. The complete theory, including Dirac's transformation theory, finally became called quantum mechanics or quantum theory, a new mathematically consistent dynamics which replaced classical physics. The labor of nearly three decades had yielded a new world dynamics. The mathematical formalism was intact and triumphed experimentally. But what did it mean? What was the interpretation of quantum mechanics, and what was it saying about physical reality? Heisenberg commented, "Contemporary science, today more than at any previous time, has been forced by nature herself to pose again the old question of the possibility of comprehending reality by mental processes, and to answer it in a slightly different way."

5. Uncertainty and Complementarity

It is wrong to think that the task of physics is to find out how Nature is. *Physics concerns what we can say about Nature.*

—NIELS BOHR

DETERMINISM—THE WORLD view that nature and our own life are completely determined from past to future—reflects the human need for certainty in an uncertain world. The projection of that need is the all-knowing God some people find in the Bible, a God who knows the past and future down to the finest detail—like a film that has already been developed. We may not have seen the film, but what it holds for us is already fixed.

Classical physics supported the world view of determinism. According to classical physics the laws of nature completely specify the past and future down to the finest detail. The universe was like a perfect clock: once we knew the position of its parts at one instant, they would be forever specified. Human beings could not, of course, know the positions and velocities of all the particles in the universe at one instant. But invoking the medieval concept of "the mind of God," we could imagine that this perfect mind knows the configuration of all the particles, past and future.

With Max Born's statistical interpretation of the de Broglie–Schrödinger wave function, physicists finally renounced the deterministic world view of nature. The world changed from having the determinism of a clock to having the contingency of a pinball machine. Physicists realized that the concept of the perfect all-knowing mind of God has no support in nature. Quantum theory

—the new theory that replaced classical physics—makes only statistical predictions. But is there a possibility that beyond quantum theory there exists a new deterministic physics, described by some kind of subquantum theory, and the all-knowing mind uses this to determine the world? According to the quantum theory this is not possible. Even an all-knowing mind must support its knowledge with experience, and once it tries to experimentally determine one physical quantity the rest of the deck of nature gets randomly shuffled again. The very act of attempting to establish determinism produces indeterminism. There is no randomness like quantum randomness. Like us, God plays dice—He, too, knows only the odds.

It is this very randomness that makes the determinist recoil. Physics, as it was conceived of for centuries, was supposed to predict precisely what can happen in nature. In the quantum theory, only probabilities are precisely determined, and the determinist finds it difficult to renounce the hope that behind quantum reality a deterministic reality exists. But in fact the quantum theory has closed the door on determinism.

The randomness at the foundation of the material world does not mean that knowledge is impossible or that physics has failed. To the contrary, the discovery of the indeterminate universe is a triumph of modern physics and opens a new vision of nature. The new quantum theory makes lots of predictions—all in agreement with experiment. But these predictions are for the distribution of events, not individual events—it is like predicting how many times a specific hand of cards gets dealt on the average. Probability distributions are causally determined, not specific events.

After Born's statistical interpretation, other physicists struggled to deepen the understanding of the new quantum theory. What knowledge of nature was possible in the framework of the new theory? For example, the mathematics of quantum theory permitted both a particlelike and a wavelike representation for the electron. But clearly these two representations were opposed and in conflict with any common-sense ideas. Is the electron a wave or a particle? Bohr, Heisenberg, and Pauli in Copenhagen and many others debated these questions for over a year. Frustration set in, but Bohr's persistent optimism kept up a spirit of inquiry. Finally, by the beginning of February 1927, Bohr was exhausted and needed a break from Heisenberg, and he took a vacation, collecting his thoughts. While on vacation, Bohr had a primary insight into the meaning of the quantum theory. Likewise Heisenberg, in the absence of Bohr but under the lash of Pauli's criticism, came to his own interpretation of the quantum theory. Bohr and Heisenberg each in his own style had come to new breakthroughs in understanding which were conceptually equivalent. Heisenberg had discovered the uncertainty principle, and Bohr had discovered the principle of complementarity. Together these two principles constituted what became known as the "Copenhagen interpretation" of quantum mechanics—an interpretation that convinced most physicists of the correctness of the new quantum theory. The Copenhagen interpretation magnificently revealed the internal consistency of the quantum theory, a consistency which was purchased at the price of renouncing the determinism and objectivity of the natural world.

Heisenberg's forte was expressing physical intuitions in precise mathematical terms. His discovery of the uncertainty relation was an example of this. It came out of the existing mathematical formalism of quantum mechanics and served profoundly to clarify the meaning of that formalism.

Heisenberg, you will recall, invented the matrix mechanics in which the physical properties of a particle such as its energy, momentum, position, and time were represented by mathematical objects called matrices, generalizations of the idea of simple numbers. Simple numbers obey the commutative law of multiplication —the result of the multiplication does not depend on the order in which it is carried out: $3 \times 6 = 6 \times 3 = 18$. But matrix multiplication *can* depend on the order in which it is carried out. For example, if A and B are matrices, then $A \times B$ does not have to equal $B \times A$.

What Heisenberg showed was that if two matrices representing different physical properties of a particle, like the matrix q for the position of the particle and the matrix p for its momentum, had the property that $p \times q$ did not equal $q \times p$, then one could not simultaneously measure both these properties of the particle with arbitrarily high precision. To illustrate this, suppose that I build an apparatus to measure the position and momentum of a single electron. The readout of the apparatus consists of two sets of numbers, one marked "position," the other "momentum." Every time I push a button the apparatus simultaneously measures the position and momentum of the electron and prints out two long numbers, the result of the measurements. For any single measurement, let's say the first one, the two numbers printed out can be as long as I like and hence highly precise. I might think that I have simultaneously measured both the position and momentum of the electron with incredible accuracy. However, to get an idea of the error or uncertainty in this first measurement, I decide to repeat the measurement and push the button again. Again two long numbers are printed out representing the position and the momentum of the electron. Remarkably, they are not exactly the same as for the first measurement, although perhaps the first several digits of each position and momentum measurement agree. By pushing the button again and again I assemble more and more such measurements. Then I can calculate the uncertainty of the position and momentum of the electron by a statistical averaging procedure over the entire set of measurements, so that the quantity denoted by Δq is the spread or uncertainty of position measurements about some average value and likewise Δp is the spread of momentum measurements about an average value. The uncertainties Δq and Δp have meaning only if one does a large set of measurements so that one can compare the differences between individual measurements. What the Heisenberg uncertainty relation asserts is that it is impossible to build an apparatus for which the uncertainties so calculated, over a large series of measurements, fail to obey the requirement that the product of the uncertainties, $(\Delta q) \times (\Delta p)$, is greater than or equal to Planck's constant, h. This is expressed mathematically by the relation:

$$(\Delta q) \times (\Delta p) \geq h$$

A similar uncertainty relation is found for the uncertainty in the energy, ΔE, of a particle and the uncertainty in the elapsed time, Δt:

$$(\Delta E) \times (\Delta t) \geq h$$

Heisenberg derived these formulas directly from the new quantum theory.

To see what these relations imply, suppose we try to measure the position of an electron with arbitrarily high accuracy. This means that our uncertainty in the position of the electron is zero, $\Delta q = 0$—we know its location exactly. But the Heisenberg uncertainty relation says that the product of Δq and Δp, the uncertainty in the momentum, must be greater than a fixed quantity, Planck's constant. If Δq is zero, this means Δp must be infinite—that is, the uncertainty in our knowledge of the momentum of the particle is infinite. Conversely, if we knew precisely that the electron was at rest, so the uncertainty in the momentum is zero, $\Delta p = 0$, then the position uncertainty, Δq, must be infinite—we have no idea where the particle is located. As Heisenberg remarked, the uncertainty in position and momentum are like "the man and the woman in the weather house. If one comes out the other goes in." Notice that if Planck's constant h were equal to zero in the real world rather than a tiny number, then we could simultaneously measure both the position and momentum of a particle, because the uncertainty relation would be $(\Delta q) \times (\Delta p) \geq 0$ and both Δq and Δp could be zero. But because Planck's constant is not zero, this is impossible.

I have always thought that wet seeds from a fresh tomato illustrate the Heisenberg relation. If you look at a tomato seed on your plate you may think that you have established both its position and the fact that it is at rest. But if you try to measure the location of the seed by pressing your finger or a spoon on it the seed will slip away. As soon as you measure its position it begins to move. A similar kind of slipperiness for real quantum particles is expressed mathematically by the Heisenberg uncertainty relations.

An important warning must be stated regarding Heisenberg's uncertainty relation: it does not apply to a single measurement on a single particle, although people often think of it that way. Heisenberg's relation is a statement about a statistical average over lots of measurements of position and momenta. As we showed, the uncertainty Δp or Δq has meaning only if you repeat measurements. Some people imagine that quantum objects like the electron are "fuzzy" because we cannot measure their position and momentum simultaneously and they therefore lack objectivity, but that way of thinking is inaccurate.

To get a feeling of what the Heisenberg relation implies for various objects, we can compare the product of the size of an object times its typical momentum to Planck's constant, h—a measure of how important quantum effects are. For a flying tennis ball, the uncertainties due to quantum theory are only one part in about ten million billion billion billion (10^{-34}). Hence a tennis ball, to a high degree of accuracy, obeys the deterministic rules of classical physics. Even for a bacterium the effects are only about one part in a billion (10^{-9}), and it really doesn't experience the quantum world either. For atoms in a crystal we are getting down to the quantum world, and the uncertainties are one part in a hundred (10^{-2}). Finally, for electrons moving in an atom the quantum uncertainties completely dominate and we have entered the true quantum world governed by the uncertainty relations and quantum mechanics.

Once I tried to imagine what I would see if I could be shrunk down to the size of atoms. I would fly around the atomic nucleus and see what it was like to be an electron. But as I came to understand the meaning of the Copenhagen interpretation of Bohr and Heisenberg I realized that the quantum theory, with its emphasis on super-realism, explicitly denies such a fantasy. I had been trying to form a mental picture of the atom based on the world of my ordinary visual experience, which obeys the laws of classical physics, and applied it precisely where quantum physics says such a picture cannot be maintained. Bohr would insist that if you want to indulge this fantasy then you must precisely specify how the shrinking down to the size of atoms is to be accomplished. Suppose that instead of shrinking down myself I build a tiny little probe that will go down into the atom and tell me what it finds. But since the probe must also be built out of atoms and particles —there isn't anything else out of which to build it—the probe, too, becomes subject to the uncertainty relations, and then we can't even visualize the probe. You see we are stuck. All we can do is perform experiments on atoms and quantum particles resulting in measurements recorded on macroscopic-sized instruments. The quantum theory describes all possible such measurements; we cannot do better. The fantasy of the shrinking person is just that—a fantasy.

Heisenberg's uncertainty relations tied in nicely with Born's previous discovery of indeterminacy, thus deepening physicists' understanding of the internal consistency of the quantum theory. Physicists found that the uncertainty relations *implied* indeterminacy. We can see this easily if we imagine a standard problem in ballistics. Suppose a gun is fired (in outer space, so we can ignore air resistance) and we know both the position of the bullet as it leaves the barrel and its momentum. Using the laws of classical physics and the knowledge of the bullet's initial position and momentum, we can precisely determine its entire future trajectory; everything is predetermined. If we consider this same problem from the standpoint of quantum theory, we are required to draw a different conclusion. The Heisenberg uncertainty relation implies that we cannot simultaneously establish both the position and momentum of the bullet at the instant it leaves the barrel. Hence to the degree that these initial measurements are uncertain, so also is the future trajectory of the bullet undetermined. All we can do is give a statistical or probabilistic description of the future trajectory of the bullet. For real bullets, like tennis balls, these quantum effects are negligible. But for electrons we are compelled to have a probablistic description of their future motion. That is why Heisenberg's uncertainty relations imply Born's indeterminacy.

While Heisenberg worked on the uncertainty relations, Bohr,

in his very different style, independently developed his own interpretation of the quantum theory. Heisenberg's approach was to use mathematics to extract the meaning of the new theory, while Bohr reflected philosophically on the nature of quantum reality. Each physicist's approach supplemented and enriched the other's, and together they constituted the Copenhagen interpretation.

Bohr wondered how we could even talk about the atomic world —it was so far removed from human experience. He struggled with this problem—how can we use ordinary language developed to cope with everyday events and objects to describe atomic events? Perhaps the logic inherent in our grammar was inadequate for this task. So Bohr focused on the problem of language in his interpretation of quantum mechanics. As he remarked, "It is wrong to think that the task of physics is to find out how Nature *is*. Physics concerns what we can say about Nature."

Bohr emphasized that when we are asking a question of nature we must also specify the experimental apparatus that we will use to determine the answer. For example, suppose we ask, "What is the position of the electron and what is its momentum?" In classical physics we do not have to take into account the fact that in answering the question—doing an experiment—we alter the state of the object. We can ignore the interaction of the apparatus and the object under investigation. For quantum objects like electrons this is no longer the case. The very act of observation changes the state of the electron.

The fact that observation can change what is being observed can be seen from examples drawn from ordinary life. The anthropologist who studies a small village isolated from modern life will by his mere presence alter village life. The object of his knowledge changes as a consequence of examination. The fact that people know they are being observed can alter their behavior.

Nature can be most passively accommodating to the quantum experimenter. If he wants to measure the position of an electron with arbitrarily great precision and sets up an apparatus to do this, no law of the quantum theory prevents a definite answer. By "position" I always mean a statistical averaging of many position measurements. The experimentalist would conclude that the electron is a particle, an object at a definite point in space. On the other hand, if he is interested in measuring the wavelength of the electron and sets up another apparatus to do this, he will also get a definite answer. Doing the experiment this way, he would conclude that the electron is a wave, not a particle. No conflict exists between the particle and the wave concept, because, as Bohr taught us, the outcome of the experiments depends on the experimental arrangement, and different experimental arrangements are needed to measure the position and the wavelength of the electron.

Now the experimenter gets persistent. He is fed up with this wave-particle, momentum-position duality nonsense and decides to settle the question once and for all by setting up an apparatus that will try to measure both the position and the momentum of the electron. Nature now becomes most stubborn, because the experimenter runs into the brick wall of the uncertainty relation. No amount of experimental technique seems to do any good, because it is a question of principle that he is up against. Why can't you measure the position and momentum simultaneously—what prevents it? Born describes it this way: "In order to measure space-coordinates and instants of time, rigid measuring rods and clocks are required. To measure momenta and energies, arrangements with movable parts are needed to take up and indicate the impact of the object. If quantum mechanics describes the interaction of the object and the measuring device, both arrangements are not possible." Born is describing that odd feature of the laws of quantum mechanics which imply that we cannot build an apparatus that measures both position and momentum simultaneously—the experimental arrangements for these two measurements are mutually exclusive. Trying simultaneously to measure both position and momentum precisely is like trying to look at the space both in front and behind your head without using a mirror. As soon as you turn to look behind you the space behind your head also turns. You cannot simultaneously see both the space in front of and behind you.

Particle and wave are what Bohr called complementary concepts, meaning they exclude one another. In the analogy between language and mathematics we gave previously, these complementary concepts are different representations of the same object. Physicists speak of the particle representation or the wave representation. Bohr's principle of complementarity asserts that there exist complementary properties of the same object of knowledge, one of which if known will exclude knowledge of the other. We may therefore describe an object like an electron in ways which are mutually exclusive—e.g., as wave or particle—without logical contradiction provided we also realize that the experimental arrangements that determine these descriptions are similarly mutually exclusive. Which experiment—and hence which description one chooses—is purely a matter of human choice.

Bohr was a philosopher and liked to extend his complementarity principle to other than problems of atomic physics. For example, "duty to society" and "family commitment" as portrayed in Sophocles' *Antigone* were complementary concepts and mutually exclusive in a moral context. As a good citizen, Antigone should consider her brother, who was killed trying to overthrow the king, a traitor. Her duty to the king and society was to renounce her brother. Yet her commitment to her family requires her to bury his body and honor his memory. Bohr later in his life thought that the principle of complementarity applied to the problem of determining the material structure of living organisms. We could either kill an organism and determine its molecular structure, in which case we would know the structure of a dead thing, or we could have a living organism but sacrifice knowledge of its structure. The experimental act of determining the structure also kills the organism. Of course, this latter idea is completely wrong, as molecular biologists have shown in establishing the molecular basis for life. I cite this example because it shows that even if you are as smart as Bohr, extending principles of science beyond their usual domain of application may lead to spurious conclusions.

We thus come to the two crucial points about quantum reality that emerge from the work of Heisenberg and Bohr, the Copenhagen interpretation. The first point is that quantum reality is statistical, not certain. Even after the experimental arrangement has been specified for measuring some quantum property, it may be necessary to repeat the precise measurement again and again, because individual precise measurements are meaningless. The microworld is given only as a statistical distribution of measurements, and these distributions can be determined by physics. The attempt to form a mental picture of the position and momentum of a single electron consistent with a series of measurements results in the "fuzzy" electron. This is a human construct attempting to fit the quantum world into the limitations of everyday sense awareness. People who engage in such constructions or try to find objective meaning in individual events are really closet determinists.

The second main point is that it is meaningless to talk about the physical properties of quantum objects without precisely specifying the experimental arrangement by which you intend to measure them. Quantum reality is in part an observer-created reality. As the physicist John Wheeler says, "No phenomenon is a *real* phenomenon until it is an *observed* phenomenon." This is radically different from the orientation of classical physics. As Max Born put it, "The generation to which Einstein, Bohr, and I belong was taught that there exists an objective physical world, which unfolds itself according to immutable laws independent of us; we are watching this process as the audience watches a play in a theater. Einstein still believes that this should be the relation between the scientific observer and his subject." But with the quantum theory, human intention influences the structure of the physical world.

In summary, the Copenhagen interpretation of the quantum theory rejected determinism, accepting instead the statistical nature of reality, and it rejected objectivity, accepting instead that material reality depended in part on how we choose to observe it. After hundreds of years the world view of classical physics fell. Here, from the very substance of the universe—the atom—the physicists learned a new lesson about reality.

During 1927, through many discussions with Heisenberg and Pauli, Bohr labored on a paper expressing his ideas about complementarity. He was going to present it first in Como at a meeting in honor of the Italian physicist Alessandro Volta and finally at the fifth Solvay conference in Brussels, where many of the leaders of physics would be present, including Einstein. Einstein and Bohr had first met in 1920, when both were already physicists of international stature. More than a professional relation of friendship emerged, and they developed a deep respect and love for one another. Bohr expressed great expectations that the complementarity principle would convince Einstein of the correctness of the quantum theory. When Bohr presented the Copenhagen interpretation of quantum mechanics at the Solvay conference, most physicists present accepted the new work for the triumph of understanding that it was. Not Einstein. He continued to pick away at Bohr's and Heisenberg's ideas, devising thought experiments that would show some flaw in the Copenhagen interpretation. Every time he thought he found a flaw, Bohr would find an error in his reasoning. Nevertheless, Einstein persisted. Finally Paul Ehrenfest said to him, "Einstein, shame on you! You are beginning to sound like the critics of your own theories of relativity. Again and again your arguments have been refuted, but instead of applying your own rule that physics must be built on measurable relationships and not preconceived notions, you continue to invent arguments based on those same preconceptions." The meeting ended with Einstein unconvinced, and this was a profound disappointment to Bohr.

Three years later at the next Solvay conference, Einstein arrived armed with a new thought experiment—the "clock in the box" experiment. Einstein imagined one had a clock in a box preset so that it would open and close a shutter very quickly on the light-tight box. Inside the box was also a gas of photons. When the shutter opened, a single photon would escape. By weighing the box before and after the shutter opened, one could determine the mass and hence energy of the escaped photon. Consequently, it was possible to determine both the energy and the time of escape of the photon with arbitrary precision. This violated the Heisenberg energy-time uncertainty relation,

$(\Delta E) \times (\Delta t) \geq h$, and hence, Einstein concluded, quantum theory must be wrong.

Bohr spent a sleepless night thinking about the problem. If Einstein's reasoning was correct, quantum mechanics must fail. By morning he discovered the flaw in Einstein's reasoning. The photon, when it escapes from the box, imparts an unknown momentum to the box, causing it to move in the gravitational field which is being used to weigh it. However, according to Einstein's own theory of general relativity, the rate of a clock depends on its position in the gravity field. Since the position of the clock is uncertain by a small amount because of the "kick" it gets when the photon escapes, so is the time it measures. Bohr showed that the thought experiment devised by Einstein did not in fact violate the uncertainty relation but rather confirmed it.

After this, Einstein never disputed the consistency of the new quantum theory. He continued to dispute that it gave a complete and objective description of nature. This objection, however, was a philosophical issue and not one of theoretical physics. The debate between Einstein and Bohr continued throughout their lives, but it was never resolved. Nor could it have been. Once the debate left the common assumption that reality is instrumentally determined, and became a difference in the appreciation of the nature of reality, there was no possibility of resolution. Bound together by a mutual love, the two titans, the last of the classical physicists and the leader of the new quantum physicists, debated to the end of their days.

By the late 1920s the interpretation of the new quantum theory was intact. A generation of young physicists grew up with it, but they were less interested in the problems of interpretation than in applications. The new theory emphasized, as never before, the paramount role of mathematics in theoretical physics. Individuals with great technical power in abstract mathematics, and the ability to apply it to physical problems, came to the fore.

The new quantum theory became the most powerful mathematical tool for the explication of natural phenomena that ever fell into human hands, an incomparable achievement in the history of science. The theory released the intellectual energy of thousands of young scientists in the industrial nations of the world. No single set of ideas has ever had a greater impact on technology, and its practical implications will continue to shape the social and political destiny of our civilization. We have made contact with new components of the cosmic code—the immutable laws of the universe—which are now programming our development. Practical devices, such as the transistor, the microchip, lasers, and cryogenic technology, have given rise to entire industries at the vanguard of technical civilization. When the history of this century is written, we shall see that political events—in spite of their immense cost in human lives and money—will not be the most influential events. Instead the main event will be the first human contact with the invisible quantum world and the subsequent biological and computer revolutions.

With the new quantum theory the basis for the periodic table of chemical elements, the nature of the chemical bond, and molecular chemistry became understood. These new theoretical developments, supported by experimental research, gave rise to modern quantum chemistry. Dirac could write in a 1929 paper on quantum mechanics, "The underlying physical laws necessary for the mathematical theory of a large part of physics and the whole of chemistry are thus completely known. . . ."

The first generation of molecular biologists were inspired by Erwin Schrödinger's book *What Is Life?* in which he argued that the genetic stability of living organisms must have a material, molecular basis. These researchers, many of them trained physicists, promoted a new attitude toward genetics and introduced the experimental methods of molecular physics foreign to most biologists of that time. This new attitude toward the problem of life culminated in the discovery of the molecular structure of DNA and RNA, the physical basis for organic reproduction. It was no accident that this discovery was made in a molecular physics laboratory, a discovery that began another revolution in its own right.

The quantum theory of solids was developed. The theory of

electrical conductivity, the band theory of solids, and the theory of magnetic materials were all an outgrowth of the new quantum mechanics. In the 1950s there were major breakthroughs in the theory of superconductivity, the phenomenon of electrical current flow without resistance at very low temperature; and of superfluidity, the motion of liquids without friction. The theory of the phase transitions of matter, such as liquid to gas or solid, was advanced.

The new quantum theory provided the theoretical apparatus for the exploration of the atomic nucleus, and nuclear physics was born. The basis for the enormous energy release in radioactive decay was understood—radioactive decay was a nonclassical process involving quantum mechanical events. Physicists knew for the first time the source of the energy of the stars, and astrophysics became a modern science.

Remarkably, the educated public did not follow these developments. The quantum theory never attracted the attention of the public as did the earlier relativity theory. For this there are several reasons. First, in the early 1930s an economic depression was going on. Second, attention to political ideologies preoccupied many intellectuals. Third, and I think most important, the abstract mathematical character of the quantum theory did not relate to immediate human experience.

The quantum theory is a theory of an instrumentally detected material reality—there is an apparatus between the human observer and the atom. Heisenberg commented, "Progress in science has been bought at the expense of the possibility of making the phenomena of nature immediately and directly comprehensible to our way of thought." Or again, "Science sacrifices more and more the possibility of making 'living' the phenomena immediately perceptible by our senses, but only lays bare the mathematical, formal nucleus of the process."

Heisenberg was interested in the contrast between Goethe, the German romantic poet and dramatist, and Newton, regarding color theory. Goethe was interested in colors as an immediate human experience, and Newton was interested in color as an abstract physical phenomenon. On an experimental, material basis one must side with Newton's conclusions. But Goethe's view —and he was one of the fathers of vitalism—speaks to the immediacy of human experience. Vitalists believe there is a special "life force" in living organisms not subject to physical laws. While this appeals to our experience, there is no material basis for it. Life depends only on how ordinary matter is organized. Life-force vitalists are rare today, but they have been replaced by those who believe that human consciousness has some special property that goes beyond the laws of physics. Such neovitalists, searching for the roots of consciousness beyond material reality, might be in for another disappointment.

Goethe was part of the romantic reaction to classical mechanics and modern science—a reaction that continues to this day. This confrontation between Goethe and Newton revealed a modern humanist critique of science that the abstract explanations of science deny the vital core of human experience. The quantum theory and the sciences that emerged from it are prime examples of such abstract explanations.

Science does not deny the reality of our immediate experience of the world; it begins there. But it does not remain there, because the basis for comprehending our experience is not given with sensual experience. Science shows us that supporting the world of sensual experience there is a conceptual order, a cosmic code which can be discovered by experiment and known by the human mind. The unity of our experience, like the unity of science, is conceptual, not sensual. That is the difference between Newton and Goethe—Newton sought universal concepts in the form of physical laws, while Goethe looked for the unity of nature in immediate experience.

Science is a response to the demand that our experience places upon us, and what we are given in return by science is a new human experience—seeing with our mind the internal logic of the cosmos. The discovery of the indeterminate universe by the quantum physicists is an example. The end of determinism meant not the end of physics but the beginning of a new vision of reality.

Here in the atomic core of matter, physicists found randomness.

But what is randomness? To examine this question we will make a little excursion off the main road to quantum reality in the next several chapters. In our excursion we will explore the chaotic universe and have a first look into the hand of the God that plays dice.

6. Randomness

It is remarkable that a science which began with the consideration of games of chance should have become the most important object of human knowledge.
—Marquis de Laplace

A few years ago I was walking through the old city of Jerusalem with an acquaintance, an archaeologist. Jerusalem is simply home to many people, but for most visitors it is a powerful, holy place. Rationality counts for little here; the symbols of faith are the real currency of this city.

In medieval times, religious people perceived Jerusalem as the center of the universe, the navel of the world where heaven and earth joined. Here at the center of the world God spoke to His prophets and the people of the Book. Jews come to worship at the wall of their ancient temple near the Holy of Holies. Christians follow the steps of their Lord in His final Passion, and Muslims worship at the third-holiest place of Islam, the Dome of the Rock, upon which the Prophet Mohammed received the Koran. God may be omnipresent, but His voice is in Jerusalem.

The archaeologist remarked that there was an ancient center of the old city marked by a Roman crossroads, which divided the city and the earth into four quadrants—the fulcrum of medieval geography. The roads had long ago disappeared, but at each corner of the crossroads had stood a Roman column which survives to the present day. As we made our way across the city to the very center of the ancient universe, my friend explained that the Roman columns stood in the interior of a modern building. Entering the building, I immediately saw the four columns. And there, between and around the columns, stood several pinball machines. Here, at the very center of the universe, was the only pinball parlor in the old city of Jerusalem. I was amazed. According to the Bible the Lord speaks only to those who are ready for His message. The prophecy was not lost—I had seen a revelation of the God who plays dice.

Major technologies often enter our civilization in an innocent and undemanding way. Some devices, which eventually become important material forces, first appear as toys. Gunpowder was first used for fireworks entertainment. The use of steam power in Hellenic Alexandria around A.D. 100 is another good example. The Greeks saw in Hero's steam wheel only a toy, a novelty, but centuries later, steam engines would be used as the motive power for the first industrial civilizations. The Alexandrian Greeks were not ready for that idea.

I think pinball machines are modern examples of such entertainment devices—they will eventually take us over. Determinists think of the universe as a huge clockwork; I think it is a pinball machine. Playing pinball requires total concentration, the right combination of skill and chance, a mastery of indeterminacy as the ball moves across the playboard and interacts with bumpers and cushions. The machine keeps score and you can cheat a little by shaking the machine, but not too much lest it tilt. It imitates life's randomness, rewards skill, and creates an ersatz reality which integrates into the human nervous system in a remarkable way. Someday such machines will be combined with art forms such as films and a completely artificial reality will be created. We are already a part of the pinball universe.

It is no accident that pinball machines—the symbol of the indeterminate universe—stand at the center of the world. The quantum theory implies that to know the world we must observe it, and in the act of observation, uncontrolled and random processes are initiated in the world. Also, Bohr's principle of complementarity implies that knowing everything at one time about the world—a requirement of determinism—is impossible because the conditions for knowing one thing necessarily exclude knowledge of others. The quantum theory means that we must renounce the determinist's dream that everything can be known. To more deeply appreciate the indeterminate universe revealed by the quantum theory, let us plunge into the world of chaos—a world first explored by mathematicians.

The human mind abhors chaos, finding order even if there is none. The ancients saw in the random patterns of the stars constellations of the figures of myth, and in the shapes of clouds animal or human figures. Tea leaves, in some cultures, foretell the future. The haruspex finds in the entrails of animals the destinies of peoples, and priests consult their gods by casting bones. The fact of natural randomness combined with the human disposition to see patterns in everything prepares the ground for heirophany—an appearance of the sacred. Some people hear the voice of God in the rushing wind, a burning bush, or a flowing stream.

Since ancient times people have been fascinated by randomness, manifested in the fall of the cards and the roll of dice. With the Enlightenment came the insight that chaos was subject to a mathematical science—and Laplace and other mathematicians discovered the laws of games of chance. At first it is not clear that there are any laws for governing chance occurrences. But if events occur at random, then they have an average pattern which is subject to certain laws. Rolling a die many times should produce each face coming up on the average once in six times, otherwise the die is "loaded." But is there a pattern in the sequence of individual rolls rather than just the average? How do we know if a sequence of events is like the individual rolls of a die—completely without meaning—or like the sequence of chemical bases on a DNA molecule, a genetic code that can specify the rules for making a human baby? Here we come to our first question: What is randomness?

In trying to answer this question it is important to distinguish between the mathematical and the physical problem of randomness. The mathematical problem is the logical one of defining what is meant by a random sequence of numbers or functions. The physical problem of randomness is determining whether actual physical events obey mathematical criteria for randomness. Clearly we cannot establish whether a sequence of natural events is truly random until we have a mathematical definition of randomness. Once we have such a definition, then we have the additional empirical problem of determining whether real events correspond to such a definition. For example, we can mathematically define a triangle in Euclid's geometry. But it is a separate, empirical question whether physical triangular configurations actually correspond to such a definition.

Here we run into the first problem: Mathematicians have never succeeded in giving a precise definition of randomness or the associated task of defining probability. If you go to a math library you will find lots of books on probability. How is it possible to have so much written about a topic that has not been precisely defined? What stands in the way of a precise definition? In part the problem of precisely defining randomness, or more specifically a random sequence of integers, is that if you succeed in giving an exact definition the sequence may no longer be random. Being able to say precisely what randomness is denies the very nature of randomness, which is utter chaos—how can you be precise about chaos?

To illustrate these difficulties, consider the problem of defining a random sequence of the integers from 0 to 9. For definiteness, consider the sequence

3141592653589793238462643383279502884197I . . .

where . . . means the sequence could go on another million digits and then stops. I just won't write them all down. This appears to be a pretty chaotic set of the ten integers, and we can apply to this set various tests of randomness. One test would be that any specific number, say 8, should appear on the average once every ten times, since there are ten different digits. So we could count the total number of 8s that appear in the sequence and divide by the length of the sequence. This ought to give a number very close to 1/10. Similarly for all the other integers. Assume this test is passed by the above sequence.

We can get more sophisticated in testing for randomness by dividing the sequence into blocks of ten integers so that we have

(3141592653) (5897932384) (6264338327) . . .

Now we can count the number of even integers that occur in each block of ten—the first block has 3 even integers, the second 4, the third 6, and so on. The number of even integers in a given block can be a number from 0 to 10. So after many, many blocks of 10, we can establish the distribution of even numbers—how many blocks have no even numbers, how many blocks have 1 even number, and so on—close to what is called a Poisson distribution. If as we increase the number of blocks the distribution actually approaches precisely the Poisson distribution, this is a further indication that the sequence is random.

We could apply other tests to the sequence as well. For example, we could use blocks of twelve integers, or any other number, to check the Poisson distribution. Suppose it passes these tests with flying colors. Can we then conclude that the sequence of integers is random? Unfortunately we can never reach such a conclusion, even if all the tests are passed, because the sequence may be specified by a rule and then it does not correspond to what we think of as random—not governed by any rule.

For example, the above sequence is just the decimal expansion of $\pi = 3.14159\ldots$, the ratio of the circumference to the diameter of a circle, which is not a random number; it is a very special number. You might object to this trick and change one of the digits, say the 100,000th, in the sequence so that the sequence isn't given exactly by π any more. But then I can change the rule with a statement that the 100,000th digit be changed, and still have a rather simple rule. The decimal expansion of π is an example of a number which is easy to specify completely in words. Actually calculating the number from its specification requires a computer program with lots of rules. For π those calculational rules are rather easy, but for other numbers they may not be.

For any finite sequence of integers, I can always find a rule that tells me exactly how to construct the sequence. But the rule may be very complicated. Andrei Kolmogorov, the great Soviet mathematician, thought he could define randomness by the criterion that if it took as long to state the rule, suitably transcribed into numbers, for the construction of the numerical sequence than the actual length of the sequence, then the sequence was "random." However, finding the construction rule for the sequence depends on human cleverness, and we can never be assured that the rule we have found is the simplest one that gives the sequence. If we did not notice that the above sequence was based on the expansion of π we would have a very difficult time finding the rule; but the rule is really very simple. A precise mathematical definition of randomness for finite sequences simply does not exist.

So there you have it: Mathematicians don't know what randomness is! But they can tell if a sequence of numbers is not random because it would flunk one of the tests for randomness. For example the sequences 33333333 . . . or 32323232 . . . , since they have a pattern, would fail the tests and are obviously not random. But even if a sequence of numbers passes all tests we cannot be certain that it is random—someone may invent a new test and it might fail. If a stock analyst sees a pattern in a sequence of prices and you see none—just randomness—we cannot decide who is right. The mathematician Mark Kac stated an amusing feature of random numbers: "A table of random numbers, once printed, requires no errata." A random number altered in a random way remains random. But we do not have an intrinsic definition of randomness. Perhaps it is impossible to give one—randomness may be absolutely undefinable.

So how do the mathematicians write all those books without defining randomness or probability? They get away with it by becoming operationalists—they give an operational definition of randomness and probability as that which obeys the theorems they derive about it. The mathematical theory of probability begins *after* probabilities have been assigned to elementary events. How probability is assigned to elementary events is not discussed, because that requires an intrinsic definition of the randomness of events—which is not known. This operational approach if applied to geometry would be like proving all sorts of theorems about triangles without actually precisely defining what is a triangle. An operational definition of "triangle" is simply the logical object that obeys all these theorems. One asks only for consistency, not for definition. You can really go very far with this approach, and that is what is in all those probability books.

This operational approach to randomness works fine most of the time. But it's possible to get into trouble, as the following example shows. Very often in computing on an electronic computer it is too time-consuming and expensive to calculate certain functions completely. Instead one calculates the functions at a set of random points and interpolates between the points, and an accurate estimate of the function can be obtained with a great saving in computer time. A similar procedure is used in opinion poll taking. It is too expensive to ask everyone his or her opinion, so a representative sample—one picked "at random"—is chosen and these are polled instead. An accurate estimate of everyone's opinion can be gotten provided the small sample is really random.

The Monte Carlo method is such a computational technique used in mathematics and physics, but it is not foolproof. Computers have in them random-number generators for the purpose of taking a random "opinion poll" of mathematical functions to speed up calculations. One method used to generate the random numbers in the computer is from the numerical solutions of algebraic equations. Such numbers pass all the usual tests of randomness, like the number for π, but they are really constructed by an algebraic rule which is programmed into the computer. They are called "pseudo-random" numbers because no one is hiding the fact that they have been constructed using a rule. Once someone did a Monte Carlo computation using these pseudo-random numbers but kept getting a nonsense answer. The reason turned out to be that the calculation he was doing involved intersecting planes in a multidimensional space which were precisely related to the solutions of the algebraic equations used in the random-number generator. For this particular calculation, the "pseudo-random" numbers generated by the computer were not at all random.

This example illustrates that what we may think is a random number really isn't—it is related to other numbers which are specified by a simple rule. How can you be sure a number is truly random? You can't—the most you can do is establish if the number is not random if it fails one test for randomness.

A remarkable feature of "random" numbers—numbers that pass all the tests—is that two such numbers may be related to one another in a nonrandom way. Suppose we consider the sequence we examined before which passes all the tests for randomness

31415926535897 . . .

and now consider a second sequence

20304815424786 . . .

which also passes all the tests for randomness. It would appear that we have two completely random sequences; however, subtracting the second sequence from the first integer by integer (with the rule that if we get a negative number we add 10 to the result) we obtain the sequence

11111111111111 . . .

which is certainly not random. This illustrates that two random sequences can be correlated—each is individually chaotic but if properly compared by using some rule (in the above example the rule was to subtract 1 from the first sequence) then a nonrandom pattern appears. Mathematicians would say that the cross-correlation of the two sequences is nonzero.

This is one way cryptographers make unbreakable codes, used for top-secret information transmission. There are two random sequences: the message, and the key held by the person who is to receive the message. Both the message and the key are completely random sequences, and no amount of cryptographic analysis by someone who holds only the message or only the key can find the message. Only by combining the two according to some rule—like combining the above two sequences—does the meaningful message appear; the information is in the cross-correlation.

The final story of randomness—utter chaos—has not yet been told to us by the mathematicians. It seems remarkable that something so fundamental for probability theory has not been defined and even more remarkable that we can go so far in mathematics lacking a definition. By simply assuming randomness exists, mathematicians assign elementary probabilities to events, and that is their starting point. But they have not captured chaos and looked it in the eye.

7. The Invisible Hand

The most important questions of life are, for the most part, really only problems of probability.
—MARQUIS DE LAPLACE

A TEACHER OF mathematics in postrevolutionary Iran began his lecture on probability theory by holding up a die which he was going to use in a demonstration. Before he could begin, an Islamic fundamentalist student cried out, "A satanic artifact!"—referring, of course, to the die. The teacher lost his job and almost his life. The notion of probability is antithetical to those interpretations of Islam which maintain that God knows everything—there is no place for chance for many religious fundamentalists.

Had the teacher been permitted to give his lecture, we can imagine what he would have told the students. He may have emphasized the application of probability theory to the real world and begun with the operational definition of probability. This kind of definition is required because we do not have an intrinsic definition of randomness—something we learned from the last chapter. We cannot determine whether or not an actual process is truly random. All we can do is to check and see if the process passes lots of tests that random processes are supposed to pass so that it is "sufficiently random." In practice this works fine, but there is always a problem in principle—we never know if someone will devise a clever new test that reveals that what we thought was random really is not.

In spite of the mathematical difficulty of defining randomness, we can take a pragmatic attitude, as did Richard von Mises. He said that the practical definition of a random process is that it is unbeatable. The practical definition works like this. Suppose that a gambling machine is built that wins if it generates random numbers. Then over the long run you cannot beat it with any strategy, and we could say that the numbers are really random for practical purposes. If there was a flaw in the machine and the numbers were not really random and some specific number came up more often, then we could use this knowledge to beat the machine. Real randomness is unbeatable. This practical definition of randomness is good for the real world. Gambling houses and insurance companies all use it. And because randomness is unbeatable and they base their business on that fact, they always win.

If we look in nature for randomness we find the best place to look for chaos is right in the atom—there is no randomness like quantum randomness. If we test processes like nuclear radioactive decays into particles, they pass all the tests for randomness. When and where an atom decays is truly random. While we can imagine a flaw in a gambling machine, physicists find no such flaw in the quantum world. Quantum randomness cannot be beat; the God that plays dice is an honest gambler. But how do we study such randomness?

Laplace and other mathematicians had the great insight that although individual random events were meaningless, the distribution of those events was not and could be the subject of an exact science—probability theory. The central idea of probability theory is the notion of a probability distribution—the assignment of probabilities to a set of related events. A simple example is coin flipping. The probability of heads is one-half and tails is one-half, and if you add up the probabilities, they add up to one. A probability of one is the same as certainty: If you flip a coin it must come up either heads or tails.

Once you have assigned the elementary probabilities, it is possible to go much further. It is possible to compute probabilities for complex events. Take a single die. The probability of a single number from 1 to 6 coming up is assigned 1/6. Now suppose we throw two dice, as is common in most dice games. The sum of the numbers on each die can run from 2 to 12. But they are not all equally probable. For example, the only way to throw a total of 2 is for each die to come up with a 1. Since the probability for each die to come up 1 is 1/6 and each event is independent, the joint probability is given by the product 1/6 × 1/6 = 1/36. On the average, throwing two dice, you will throw a 2 only once every 36 rolls.

You can throw a 3 in two different ways. The first die comes up 1 and the second die 2. This event also has a probability of 1/6 × 1/6 = 1/36. However, one could also achieve a total of 3 by the first die coming up 2 and the second die 1. This also has a probability of 1/36. So the total probability of rolling a 3 is 1/36 + 1/36 = 1/18. Proceeding in this way of counting options, one can make a table of how the probability is distributed over the various throws.

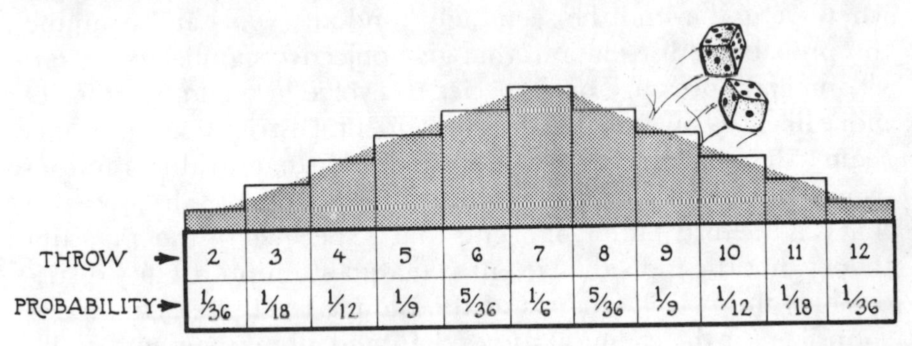

The probability of different throws from a pair of dice. It seems as if there is a "force" or an "invisible hand" which makes the 7 come up most often, but actually this is just a consequence of mathematical probabilities.

If you had to bet on a single number's coming up, the number to bet on is 7; it has the highest probability. If you had to bet on five out of the eleven numbers coming up, then bet on 5, 6, 7, 8, and 9 because they come up two out of three times; the remaining six numbers, 2, 3, 4, 10, 11, and 12, come up only one out of three times. What tells these numbers to turn up with these frequencies? We see that this probability distribution is just a consequence of mathematical combinatorics—adding up the different combinations by which a specific throw can be achieved. But it seems as if there is an "invisible hand" that pushes for the 7 more often than the other numbers. The remarkable feature of probability distributions for real events is that the distribution isn't material—yet it is manifested as a kind of invisible force on material things like dice.

Probability distributions of real events like those for throwing dice are part of the invisible world. The distributions are invisible not because like atoms they are materially small but because they are not material at all. What is visible is individual material events like the throw of dice. Probability distributions are like invisible hands that do not touch. A good example is the slow, invisible process of biological evolution. This process becomes real only when we go beyond the seemingly random events and examine the probability distributions that give objective significance to environmental pressure on a species to evolve into another species more likely to survive in that environment. Distributions of events seem to have an objectivity not possessed by an individual random event. In the microscopic world of atoms we have already seen that it is the distribution of events that is specified by the quantum theory, not the individual event. The quantum probability distributions, the invisible hands at the atomic level, are actually responsible for the chemical forces that bind atoms together.

We might imagine that probability distributions, because they have some kind of objectivity, have an existence independent of the individual event. This error can result in thinking that the distribution "causes" the events to fall into a specific pattern. This is secular fatalism—the belief that probability distributions influence the outcome of single events. But this is "backward" reasoning, because it is the single events which establish the distribution, not the other way around. By introducing a nonrandom element, an element of organization on the level of individual events, one changes the probability distribution. If you load dice they will fall differently.

While the invisibility and objectivity of distributions is amazing, another remarkable feature of probability distributions is their stability, whether they are distributions of atomic motions in matter, chemical reactions, or biological and social events. A stable distribution, one which does not change with time, is called an equilibrium distribution. The probability distributions of dice throws we do not expect to change in time, because the dice are not subject to temporal forces. But what about the probability of breaking a leg in a ski accident at a specific resort, winter season after season? How does one account for the stability of that probability over long periods of time? The stability of an equilibrium distribution is a consequence of the fact that the individual event is random and independent of similar events. Individual chaos implies collective determinism.

My favorite example of the statistical stability of distributions is dog bites in a large city. If people get bitten by dogs and report to doctors or hospitals for treatment the event is recorded. Over several years the reports number 68, 70, 64, 66, 71, an average of about 68 per year. Why is this number so stable? Why doesn't a year pass with only 5 dog bites or as many as 500? Is there some mysterious spirit that affects these dogs and people each year so that almost exactly 68 people are bitten each year without fail? The spirit that affects them is the God that plays dice. Because the events are random and independent, the distribution is stable. Only if one introduces a nonrandom element, such as a law requiring people to own less aggressive breeds of dogs, can the equilibrium distribution be altered.

This stability of probability distributions can produce a sense of pessimism regarding personal freedom, as was described by Jacques Monod, the French molecular biologist, in his account of biological evolution, *Chance and Necessity*. Monod knew that the evolution of a species is collectively determined because of the randomness of events for the individuals of the species. But what does this fact imply for human liberty? You may think you exercise your freedom by promoting a specific political opinion or deciding to wear blue shoes, but in fact, your actions are just part of a probability distribution. In French society of the 1970s there was a definite probability that a person would be on the left or right politically. What is perceived as freedom by the individual is thus necessity from a collective viewpoint. The die when it is thrown may "think" it has freedom, but whatever it does it is part of a probability distribution; it is being influenced by the invisible hand. We cannot act without being part of a distribution—it is like being in an invisible prison held by invisible hands. Even the very act of trying to escape is again part of a new distribution, a new prison. Perhaps this is why real creativity is so difficult—thousands of invisible hands hold us to our conventional acts and ideas.

There seem to be two kinds of people in this world, who in their extreme forms are those who see everything in the world as caused and meaningful and those who believe that God plays dice and that truly random occurrences happen. The determinists find it difficult to accept the random nature of some events because their minds are repelled by chaos, the uncaused. Indeed, the natural instinct of most people is to seek security in determinism and to find "reasons" for why things happen, no matter how farfetched the reason may be. Nothing brings out the distinction between the determinist and the acausalist so much as the confrontation with serious illness in another person or even oneself. The determinist will seek meaning and a cause for such illness—somehow the person has "sinned" to bring this affliction upon himself. In fact the illness may have no cause at all and, like a radioactive atomic decay, it just happens.

We are very sensitive to shifts in the distribution of our life's events. There are times when one gets the feeling that nothing is going right in one's life, or alternatively, that everything is turning out positive. Often these feelings are reactions to fluctuations in an equilibrium distribution of life-enhancing and life-diminishing events. If on the same day one loses a job, the car won't start, and a friend dies, then it is hard not to allow these events to affect one's disposition. Such fluctuations may not be very important, however. What is important is to take action when there are major material changes in life that affect the equilibrium distribution itself.

A great concern of people is their expectation for a long, healthy life. The probability distribution for life expectancy can be given for individuals of a given age, sex, and income level. However, we can change the probability associated with our life expectation by eliminating life-destroying personal habits and introducing a routine of physical exercise. That has a real effect on the distribution. But every distribution has a tail. There is a finite chance that a person who promotes his or her health will have a short life—like the small probability of rolling the dice and coming up with "snake eyes," a 2.

One can change probability distributions by introducing nonrandom forces and constraints. But is it possible to reduce probability to absolute certainty? Can one establish some event with absolute certainty? In practice it is clear that we can arrange conditions so that some events must occur, at least with a probability so close to unity as to appear certain. But if one asks this question as a matter of principle, the answer is no—there is no absolute certainty for events. The reason for this lies ultimately in the statistical interpretation of quantum mechanics. If we mean by minimally specifying an event, its time of occurrence, and the energy change that signals the event, then the Heisenberg uncertainty principle precludes an absolute determination—the final joke of the God that plays dice.

Every distinct event can be part of a probability distribution and is subject to the invisible hand. But different sets of events may not be independent—the push of different invisible hands may be correlated. Consider the probability for having heart attacks in New York City and Boston on a specific day. We may think that such events are independent because the cities are apart, but they are not. If more heart attacks occur in Boston, more also occur in New York City. By examining this we find, of course, that if the temperature is high in one city—something that makes heart attacks more likely—then it is also likely to be high in the other city. Heart attacks are related to emotional stress, which runs high on weekends, and if it is a weekend in one city it is also a weekend in the other. Events that superficially appear independent are in fact related. Mathematicians would say that the two probability distributions for heart attacks in New York and Boston are cross-correlated.

In our discussion of randomness we pointed out that two random sequences, like sequences of numbers, if compared could have a nonrandom cross-correlation. Chaos can be related to chaos in an ordered way. A beautiful illustration of this idea of cross-correlation of random functions is Bella Julesz's random stereograms. The stereopticon was a Victorian version of the slide projector, a device for viewing pictures, usually photographs of tourist attractions. The slides consisted of ordinary photographs, side by side, almost identical because they were taken through two lenses somewhat displaced from each other the way our eyes are displaced from each other. The stereopticon splits the two images so that one image goes into the right eye, the other into the left. The result is that the brain fuses the two images into a three-dimensional image of a scene.

Julesz's idea was to use completely random sequences of spots, like spots on a television screen, for the two images. However, the cross-correlation of the two random pictures was not random but had a discernible pattern. These two pictures, when used in a stereopticon, put completely random information into the right eye and the left eye individually. But, amazingly, the brain is sensitive to the cross-correlation. Remarkably, when at first you look at a random stereogram you don't see anything except random patterns of dots. After a few minutes the pattern, or cross-correlation, emerges. What you see is a three-dimensional image, and it is undeniable. The brain has cross-correlated a large amount of random information. There is no way you could see the image as three-dimensional unless the brain was actually correlating the random information going into the right eye with that going into the left. If you return to the same random stereogram after waiting several weeks, the intelligible pattern appears almost immediately. Somehow the human brain can remember the cross-correlation of tens of thousands of bits of random information. I think it would be interesting to know how the brain does this—a problem in neurophysiology. What is interesting for us in the example of random stereograms is that they make the abstract concept of the cross-correlation of random functions a lived experience. Later, when we return to examining the quantum theory, we will use this notion of cross-correlation.

We have been discussing probability distributions such as the probabilities for throws of dice or dog bites in a city. But are these invisible hands real; are they objective? A good gambler knowing nothing about probability theory may have a sense of the odds almost like that of a material presence. Perhaps there exists a social being like Hobbes's Leviathan—a symbol of the collective social consciousness—for whom the distribution of human events has reality the way tables and chairs have reality for us. If such a collective consciousness exists, then I have no idea how to actually establish its existence. Those who make claims for a collective consciousness such as "the will of the people" are usually using it to serve their social or political opinions.

One can debate the question of the objective nature of the distribution of human and social events for a long time. Our purpose, however, is to examine the concept of probability distribution for our picture of material reality as given by the quantum theory. Reality in the quantum theory is statistical; the invisible hands are touching everything in the microworld. As Heisenberg remarked, ". . . in the experiments about atomic events we have to do with things and facts, with phenomena that are just as real as any phenomena in daily life. But the atoms or the elementary particles are not as real; they form a world of potentialities or possibilities rather than one of things or facts."

But before returning to the problem of quantum reality we will continue for one more chapter our excursion into the nature of randomness and probability. If we now turn to examining the probability distribution of the motions of molecules in a gas, then, unlike the distribution of heart attacks or dog bites, the touch of the invisible hand can be directly felt. How that is accomplished is quite remarkable and is the subject of the next chapter.

8. Statistical Mechanics

A good many times I have been present at gatherings of people . . . who have . . . been expressing their incredulity at the illiteracy of scientists. Once or twice I have been provoked and asked [if they] could describe the Second Law of Thermodynamics. The response was cold . . . also negative. Yet I was asking something which is the scientific equivalent of: Have you read a work of Shakespeare's?

—C. P. SNOW, *The Two Cultures*

WHAT IS A gas? For centuries people thought that gases were a continuous material medium, because that is how they appear. But in fact gases actually consist of billions upon billions of molecules rapidly flying in straight paths until they hit each other or a wall—a mad circus of particles all bumping into each other. How can we study such a vast collection of particles in rapid chaotic motion?

First of all, we can measure definite properties of a gas such as its temperature, pressure, and total volume. We don't have to know that a gas consists of lots of molecules to measure these properties; all we need is a thermometer or a pressure gauge. Such quantitative properties of a gas as temperature or pressure are called macroscopic variables, because they describe the bulk properties of a gas. Physicists discovered the thermodynamic laws that such macroscopic variables obeyed. An example is the perfect-gas law, which states that the product of the pressure times the volume of a gas is proportional to its temperature. What is remarkable is that with a rather small number of such macroscopic variables we can give a complete description of how the bulk properties of matter change. Knowing how all those particles fly around isn't needed. Physicists understood these thermodynamic laws by the middle of the nineteenth century.

While these thermodynamic laws were being discovered, the world view of classical physics and Newton's laws still provided the fundamental understanding of reality. Physicists asked if the thermodynamic laws for gases and other matter could be derived from Newton's laws and thereby be put on a more fundamental basis. Some physicists guessed that gases actually consisted of flying particles and believed that each such particle obeyed Newton's laws of motion and thus its trajectory was completely determined. They theorized that if we only had the perfect mind of God, we could track each individual particle, solve the equations for the motions of all the billions of particles, and in this way determine everything about the gas. For mortals this is an impossible task—and for God, too, if we realize (as was not realized in the nineteenth century) that the quantum theory is right and the trajectory is indeterminate. So we have a problem: How can we deduce from the laws of motion for microscopic particles the macroscopic laws of thermodynamics? How can you go from the knowledge of particles to the gas we observe? The solution of this problem was one of the great accomplishments of nineteenth-century physics. If we closely examine how physicists solved this problem it will teach us something amazing about the relation of human experience to the microworld of atomic particles—our experience is not actually reducible to atomic events.

The solution of this problem was primarily the accomplishment of the Scottish physicist James Maxwell, who was famous for his laws of electromagnetism; Ludwig Boltzmann, a physicist who bore the burden of convincing his contemporaries of the correctness of his great accomplishment; and J. Willard Gibbs, an obscure American genius working alone at Yale University. These men created statistical mechanics. The basic hypothesis of statistical mechanics is that all matter is made of particles—these can be identified as atoms or molecules—whose motion obeys the mechanical laws of classical, deterministic physics. In practice it is impossible to apply the mechanical laws in detail to each particle, since there are so many particles in a macroscopic piece of matter. The main idea of Maxwell, Boltzmann, and Gibbs was to apply statistical methods so that the probability distribution of particle motions is determined rather than the motion of each individual particle. For example, a gas such as the air around us consists of billions of particles all flying around until they hit each other or collide with some obstacle like a wall. According to classical physics, each particle moves in a determined way, but it is impossible to track all individual particles. However, physicists deduced from Newton's laws statistical properties of all the particles, such as their distribution of velocities or the average time between collisions. It would be like conducting a poll of all the particles asking each one what its velocity is and then establishing a distribution for the velocities. Obtaining the thermodynamic laws as statements about the distributions of particle motions is the basic accomplishment of statistical mechanics.

If you blow up a balloon, you have filled it with billions of particles of air. The pressure on the inside of the balloon is due to these particles hitting the inside of the elastic surface and transferring momentum to it. We can feel the pressure by squeezing the balloon. Pressure is a macroscopic property of the gas, but it can be derived from the mechanical laws of particles hitting the physical surface which contains the gas. Similarly, the temperature of a gas, another macroscopic variable, can be related to the average energy of motion of the particles. There is no concept of temperature for a single particle; only if you have lots of particles does temperature emerge as a collective property. Temperature and pressure are average properties of the distribution of motions of a collection of particles. Unlike the probability distributions we encountered in our discussions of dice throwing or human and social events, the distributions in statistical mechanics are directly perceptible as the temperature or pressure of a gas —invisible hands which we can directly feel.

In the course of discovering the thermodynamic laws, physicists discovered yet another macroscopic variable which described a bulk property of matter—entropy. Entropy is a quantitative measure of how disorganized a physical system is, a measure of its messiness. This concept of entropy is extremely important for our understanding of the relation of the microworld to the world of our human experience. Have you ever noticed how hard it is to keep things neat and organized? Your frustration is not an accident but a consequence of one of the fundamental laws of thermodynamics—entropy or messiness always increases for a closed physical system. You are fighting the second law of thermodynamics.

To illustrate entropy increase, take a glass jar and fill it up a quarter of the way with salt. Then add granulated pepper until it is half full. There is a black layer on top of a white layer—an improbable configuration of all the particles. This configuration has relatively low entropy because it is highly organized and not messy. Now shake the jar vigorously. The result is a gray mixture, a disorganized configuration of the salt and pepper. If you keep shaking it is very unlikely that the original configuration will ever return. Not in millions of years of shaking will it return. The entropy or disorganization of the system has permanently increased. This illustrates the second law of thermodynamics: For any closed system the entropy always increases. A system will always change from a less probable configuration (black pepper on top of white salt) to a more probable configuration (gray mixture). It is important that the system be closed for the law of entropy increase to apply. If I open the jar of the gray mixture and carefully separate the salt and pepper I can recreate the original configuration.

When I was a college freshman my roommate and I decided we were tired of cleaning up the single room we shared and allowed it to go to a state of "maximum entropy." The room became a complete mess, much to our satisfaction. If we moved anything we couldn't help cleaning it up, but the next move would recreate the mess. The cleaning problem was solved, but another problem arose—we couldn't find anything. Several minutes would be spent looking for the things we needed. Eventually we decided the state of maximum entropy was not saving us time and effort and we returned to a more conventional life-style.

The law of entropy increase is manifested all about us. The

phenomenon of material deterioration is an example. Everything eventually falls apart—buildings crumble and fall into ruin; we age; fruit rots. The law of entropy increase may apply to the universe as a whole because the universe may be a closed system. Eventually it too may fall into ruin, a "heat death" in which the stars burn out and matter is scattered over the endless reaches of space—a mess with no one to straighten it out. It would be a bleak, unhappy end of time.

The law of entropy increase is interesting, because it is a fundamental law and yet it is statistical in character. Really the law of entropy increase follows because a highly organized configuration is improbable compared to a disorganized one and it is more likely for a state of nature to go from an improbable configuration to a highly probable one. The invisible hands are always creating a mess. Let us examine this more closely. Suppose we have two closed vessels A and B, both of equal volume, connected by a pipe with a closed valve. Vessel A is filled with a gas and the other, B, is empty. If we open the valve the gas rushes out of the full vessel into the empty one until an equilibrium situation is reached with equal pressure in A and B.

The gas moves in accord with the law of the increase of entropy. Once the valve is open, it is an extremely improbable configuration that all the gas remains in A and therefore it rushes into B until maximum entropy is achieved. This equal pressure configuration is simply the more probable one (like the gray mixture of salt and pepper), and that is why it happens.

From the viewpoint of the mechanical motions of the individual gas particles—the microscopic description—the situation is seen entirely differently. It is true that a gas particle will move from a high-pressure vessel A to a region of low pressure in vessel B. But since the motion of the particles is random, it is equally likely that a particle that has traveled into B will wander back into A. The mathematician Poincaré had shown, using the laws of classical physics, that a system like this one will eventually return arbitrarily close to its original configuration with all the particles in vessel A again. Does this behavior contradict the law of entropy increase? It certainly seems to.

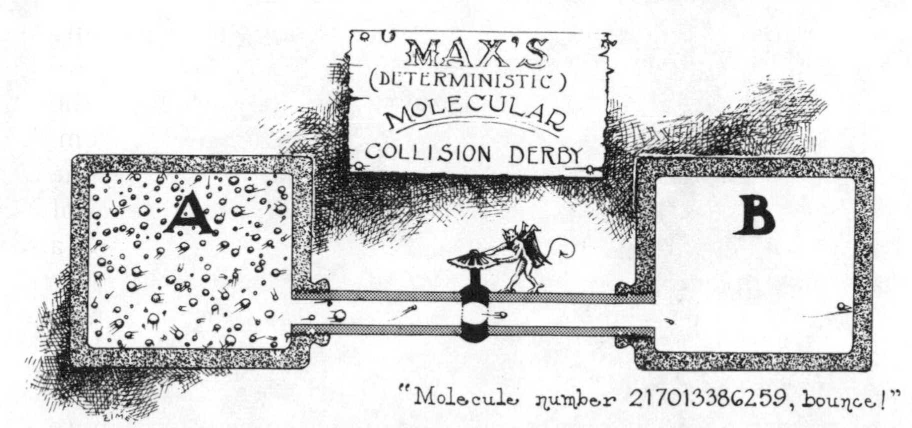

Two gas vessels, A and B, with a friendly demon opening the valve. From our viewpoint the gas flows from the full vessel A to the empty vessel B because the pressure is greatest in A. But according to statistical mechanics and the second law of thermodynamics—the law of entropy increase—this behavior is supposed to be just a consequence of probability. How then is it possible for the gas to flow into B if it is as probable for a molecule to go from A to B as to wander back from B into A?

To clarify this problem the two physicists Paul and Tatiana Ehrenfest made a simple mathematical model that imitated the statistical mechanics of the gas flowing from A to B. Imagine two dogs called A and B who represent the two gas vessels and dog A has fleas, a total of N of them, while dog B has none. Each of the fleas has its own number from 1 to N. The fleas represent the gas particles in this model. Near the two dogs is a pail containing N small balls, each with a number from 1 to N on it. The fleas can jump from dog to dog, but they have to wait until someone reaches in the pail and pulls out a numbered ball. When the number is called—"Flea number 86, jump!"—that flea jumps from whichever dog it is on to the other one. The ball is returned to the pail and mixed around and then another ball is picked, a process which can be carried out indefinitely. Notice that each flea is as likely to remain on whichever dog it is on as jump to the other one, just like the gas particles in the vessels A and B.

The Ehrenfests illustrated the statistical nature of the second law of thermodynamics by the two dogs A and B representing the gas vessels A and B. Dog A starts out with all the fleas, but as their number is called each flea jumps from whichever dog it is on to the other. Soon there is an approximately equal number of fleas on each dog—an equilibrium configuration around which there are small statistical fluctuations.

At first all the fleas are on dog A, and so most fleas jump to B as their number is called. But soon there are almost an equal number on B and it becomes as likely for a flea to jump from B to A as from A to B. This corresponds to the equilibrium configuration similar to that for the gas in the two vessels. Suppose N = 100 fleas. Then on the average there are 50 fleas on each dog. But in general there are fluctuations. Sometimes there will be 43 fleas on A and 57 on B. But for most of the time the number of fleas on either dog will be very close to 50.

It is possible, however, that after many drawings and callings of numbers a fluctuation will become very large and all the fleas will end up on one dog. This is extremely improbable for even 100 fleas. But suppose there were only 4 fleas. Then from time to time it is quite likely that all the fleas will end up on one dog.

This simple model beautifully illustrates the statistical nature of the second law of thermodynamics—the increase of entropy. The law makes sense only if we have a large number of particles so that we can speak of a probability distribution, an averaging of the motion of lots of particles. The invisible hand can exist only as a collective aspect of lots of individual particles or events. For a small number of fleas or gas particles the law is really inapplicable. One can go from an equal number of fleas on each dog to all fleas on one dog fairly often. But for real gases the number of particles is in the trillions of trillions and the probability that the original configuration will be achieved is once in billions of billions of lifetimes of our universe. It is completely improbable that a large fluctuation involving such a huge number of particles can occur. It is a safe bet, almost certain, that entropy will never decrease for a closed system with a large number of particles. The law of entropy increase is statistical, not absolutely certain.

It may seem that the God that plays dice has already entered our description of material reality because we have resorted to a statistical description of gas. But this is not so. The laws of classical physics from which the laws of thermodynamics like entropy increase were derived are still completely deterministic. They were discovered by physicists who were committed to deterministic physics. We mortals may not know how to calculate the motion of all the particles in a gas. This, however, is only a practical problem, not one of principle. Mortals must play dice and use statistical methods to find the behavior of thermodynamic systems like gases. But for the mind of God that can know the motion of every particle, no statistical method need be used—God does not play dice in classical physics. He can know exactly every fluctuation in a gas.

The statistical character of the law of entropy increase is but one aspect of this law. A very remarkable further feature of the second law of thermodynamics is that it cannot be derived from just the classical laws of motion. This feature seems to contradict the very enterprise of statistical mechanics, which was deriving thermodynamic laws from Newton's laws. But if we examine this feature it will actually deepen our understanding of the second law and the relation of physics to human experience. How are we certain this law cannot in fact be obtained from just the laws of motion of all the individual particles? This is rather easy to see.

The microscopic description of a physical system in terms of the motion of individual particles is given by Newton's laws of motion, which is our starting point. These laws of motion make no distinction between past and future; from the standpoint of the microscopic world, time can literally have either direction. An atom knows nothing about aging—it is how atoms and molecules are organized that determines age. Irreversible time, aging, the rotting of fruit are all illusions from the standpoint of microphysics. But the law of entropy increase in time gives time an arrow, a direction which distinguishes the past and the future. If a rotting fruit became whole, in violation of the law of increase of entropy, it would be like time running backward. The microscopic laws are therefore said to be time reversal invariant, and the macroscopic laws, like the law of entropy increase, are not. It is therefore impossible to derive the second law of thermodynamics, which is a law for the macroscopic variable entropy, from only the laws of Newton's mechanics. No amount of mathematical sleight of hand can get you from one set of laws to another. It is clearly necessary to make an additional assumption if we are going to derive the law of entropy increase from Newton's laws. This assumption I would like to discuss, because it tells us something about the relation of the fundamental laws of the microcosmos to our immediate experience.

To illustrate this additional assumption, let me suppose I have a remarkable motion-picture camera with a zoom lens that can focus on everything from microscopic to macroscopic objects. I will take a motion picture of a person smoking a pipe, slowly zooming the lens from the microworld to the macroworld. The film is then shown on a screen.

At first we see only the microworld of the particles of air and smoke bouncing around and hitting each other. The particles all obey Newton's laws of motion. If I were to run the projector backward, all the particles would reverse their motion on the screen. But qualitatively this motion is the same as before—it is just a mess of particles bouncing around. We cannot determine the direction of time from this microscopic view because Newton's laws don't distinguish the past from the future.

Now let us imagine that the zoom lens takes in more visual area. When we cease to see the individual particles of air and smoke bouncing around and hitting each other, then we are losing information. Instead what we begin to see, at that stage, are the wisps of smoke themselves—the *average* distribution of the smoke particles moving in time. If we ran the projector backward it might still be hard at this stage to tell the direction of time from what we see on the screen. But we would get suspicious that the film was being shown backward if we saw wisps of smoke condensing instead of expanding, because that seems improbable. What has happened is that by washing out the microscopic information in favor of macroscopic information—eliminating the individual motion of particles in favor of the average—we have slipped in the extra assumption that connects Newton's mechanics to thermodynamics. There is no way to lose detailed microscopic information without increasing entropy. Our averaging over the microworld description, which washes out the details, necessarily violates the time-reversal character of that microworld. It could be said that it is our demand itself—the demand that there exist a macroworld description, that there be a meaningful averaging of microworld information—that introduces the arrow of time.

At what point in the zoom-lens sequence do we recognize that the film is being run backward? If we now zoom out and take in the full macroworld we see a person inhaling smoke from the air and reconstituting it into tobacco in the pipe. There is no question now that the film is being shown backward. The point in the

Neils Bohr smoking his pipe. Imagine this to be a motion picture. If we look at the smoke close up all we see are particles bouncing around —like at the end of the trail of smoke—and we cannot tell the direction of time. If we take in more visual area we begin to see wisps of smoke and, perhaps, if the motion picture is shown backwards, we can tell the direction of time. Finally, the motion picture of the full view clearly has a time's arrow because if smoke is going into the pipe we know the picture is being shown backwards. Notice the vase with flowers. Or is it two profiles? This is a visual metaphor for the observer-created reality and the principle of complementarity.

zoom-lens sequence that we recognize time's arrow is pointing backward is the point that we recognize events that are impossible or unlikely. It is we who recognize the pattern, and it is we who impose the macroscopic description of physical reality, a reality which does not apply to the microworld.

We see that we can draw a line between the microworld and the macroworld of human experience—they are qualitatively distinct descriptions of material reality. Our minds and bodies respond to the thermodynamic macrovariables, which are the distributions of the microscopic motions. I feel hot or cold, a definite temperature, not the bombardment of billions of particles on the surface of my skin. I grow old and life is full of risks which have meaning only because some decisions are irreversible—I cannot go back in time. Yet from the microscopic viewpoint this is all an illusion.

The fact that we cannot logically proceed from a microworld to macroworld description without introducing a new assumption has implications for the philosophy of material reductionalism. In its crudest form, material reductionalism maintains that there is a series of levels. At the bottom level are the subatomic particles, and from these the chemical properties of atoms and molecules are obtained. Molecules form living and nonliving things, and from the behavior of molecules and cells it is possible to determine the behavior of individual humans. They in turn establish a social order and institutions. Finally at the top level of the ladder are historical events. The claim is that in principle, history is materially reducible to subatomic events.

But it is clear from our discussion of how the arrow of time is established that such a reduction is impossible even from the level of macroscopic objects to atoms. A meaningful macroworld description involves an averaging that washes out information of the microworld, and it is we who average. We can also see that historical events like a social revolution are not even reducible to individual actions on the part of humans. Two people observing a social revolution may agree completely about the events on the individual level—what was said by whom and where. But these same two people may disagree completely as to the meaning of the historical event. They see different patterns and bring their own experience to bear on the interpretation.

While we may disagree as to the patterns we see in social events, we all seem to agree on the direction of time. Physicists can rigorously derive the laws of thermodynamics by introducing the averaging over lots of microworld events, a procedure which is mathematically consistent. But why do human beings perceive the world in terms of that mathematical averaging procedure? Who told us to do that?

I think the answer lies in the process of biological evolution. There is evolutionary pressure toward the development of organisms that do not respond to individual atomic events but develop receptors that measure the properties of distributions like temperature and pressure. There was selective value in losing detailed information in favor of relevant information. Such organisms were the first pattern recognizers, and they were evolutionarily successful. Distributions, the invisible hands, acquire reality through the selective process.

Richard Dawkins has written a wonderful book, *The Selfish Gene,* which describes how biological and behavioral patterns of humans and animals can be quantitatively accounted for by assuming that specific genes—collections of protein molecules in a DNA or RNA chain—are selfish and just want to survive for all time. Genes, in their evolutionary struggle for survival, have created versatile bodies around themselves—our bodies—and civilization is just the way these clever molecules perpetuate themselves. This seems a bit much—perhaps more than the proponents of this viewpoint would themselves maintain—and indeed it is just the old philosophy of material reductionalism applied to biology. A friend of mine, a director of a molecular biology laboratory, was asked by his students about Dawkins's viewpoint. Was Dawkins really right? The next day my friend responded to his students and said that Dawkins had it all wrong. Genes were not selfish; it was in fact a specific chemical bond in the gene that was selfishly trying to perpetuate itself. Genes were just the playthings of this chemical bond, the same way we are the playthings of our genes —a *reductio ad absurdum.*

No doubt animal and human behavior is materially conditioned by the microscopic organization of genes just as the increase of entropy is materially conditioned by the motion of molecules. The evidence for that conditioning is clearly described in Dawkins's book. But it is not possible to reduce all human behavior to just microscopic genes. Genes just do what they are told to do by the laws of chemistry. The point is that genes can't be selfish but human beings can.

The microworld is indifferent to success or failure, selfishness or altruism. But nature has produced organisms that know these differences and respond to patterns that are meaningless from the standpoint of the microworld. The human world, although materially supported by the microworld, exists for its own sake. Civilization reflects a pattern of human existence which is consistent with the fundamental laws of the microworld but cannot be derived from them.

Human beings are pattern-recognizing animals *par excellence.* We can perceive distributions where other animals know only individual events. We have responded to the pattern of the whole world, a set of atomic sensations, by the creation of language. Indeed, symbolic representation is our highest organizational capability. And the possibility for language rests in our pattern-recognizing ability to attribute meaning to many individual occurrences as objective distributions. We are always talking about those invisible hands—the economy or the state of the nation—which influence our lives.

Is there meaning in random events? Is there a pattern? Synchronicity refers to the psychological phenomenon of attributing a pattern, perhaps on an unconscious level, to different random events. The leaf falls as the lovers part. Sticks are thrown to determine a pattern of the *I Ching*—the Chinese *Book of Changes.* How seductive it is to think that individual events are not random but have a cause. The mind, on some level, cannot accept the purely coincidental and seeks a meaningful pattern. Evidently that behavior has selective value in the development of our species, because we are open to recognizing patterns in any configuration, even if it is random.

It is the possibility of error that opens the door to successful change. Errors in the transmission of genetic information from generation to generation promote the evolutionary process.

In the old classical physics, even errors like those that produce mutations are completely determined in principle. Even genetic changes governing the future of evolution are knowable to an all-knowing mind. But with the quantum theory this classical picture of reality was overthrown and replaced by the indeterminate universe. Spontaneous change, indeterminate even in the perfect mind of God, occurs—a few random changes in a DNA chain producing a successful mutant. That is why the indeterminism of the quantum theory is so important for our picture of reality: In principle there is a material basis for the freedom of human consciousness and the evolution of the species.

Nature knows nothing of imperfection; imperfection is a human perception of nature. Inasmuch as we are part of nature we are also perfect; it is our humanity that is imperfect. And, ironically, because of our capacity for imperfection and error we are free beings—a freedom that no stone or animal can enjoy. Without the possibility of error and real indeterminacy implied by the quantum theory, human liberty is meaningless.

The God that plays dice has set us free.

9. Making Waves

I think it is safe to say that no one understands quantum mechanics. Do not keep saying to yourself, if you can possibly avoid it, "But how can it be like that?" because you will go "down the drain" into a blind alley from which nobody has yet escaped. Nobody knows how it can be like that.

—RICHARD FEYNMAN

WE HAVE COMPLETED our excursion into the nature of randomness, probability, and statistical mechanics and learned that although chaos reigns at the microscopic level, mathematicians are not even sure what chaos is. Further, the naive program of material reductionalism—reducing human experience to atomic events—cannot be carried out. But now we must return to our original task—finding out the nature of quantum reality from the quantum theory.

Let us imagine that we are a group of pilgrims on a journey which every thinking person must take once in his or her life. It is like the hajj—the journey every faithful Muslim makes once to the holy city of Mecca—or the pilgrimage of Chaucer's characters to Canterbury Cathedral. But our journey is not to a holy city; we must content ourselves with wandering along the road to quantum reality not certain where it will end. We are explorers and must be open to every new experience. Along the way, we will encounter various experiments and illustrations designed to expose aspects of the quantum reality we are seeking. The first experiments we examine are several "thought experiments" with waves—experiments which could actually be done but are here simply intended to sharply illustrate the weird properties of the quantum world.

The Copenhagen interpretation of the new quantum theory ended the classical idea of objectivity—the idea that the world has a definite state of existence independent of our observing it. This contrasts with our ordinary experience of the world, which supports the classical view of objectivity that the world goes on even if we do not perceive it. Wake up in the morning and the world exists much as you left it. But the Copenhagen interpretation maintains that if we look closely at the world—at the level of atoms—then its actual state of existence depends in part on how we observe it and what we choose to see. We will examine how Born's interpretation of the Schrödinger waves—the indeterminism of the quantum world—implies that objective reality must be replaced by the observer-created reality. The atomic world simply does not exist in a definite state until we actually set up an apparatus and observe it. The Copenhagen view is super-realistic—no fantasies or rationalizations about material reality are allowed.

Recall that de Broglie and Schrödinger thought that the waves in quantum theory were matter waves of some kind. But Max Born realized that these waves have nothing at all to do with any kind of material wave like an ocean wave. His statistical interpretation was that they were waves of probability for finding particles at a specific point.

Born offered an illustration to describe the difference between a matter wave and a probability wave—Max Born's machine gun. If you ignite a large quantity of gunpowder and you are standing right next to it the explosion kills you. But should you stand about 100 yards away the shock wave of the explosion reaches you but you are left unharmed. The shock wave is a real matter wave, and its influence diminishes with distance. Now take the same quantity of gunpowder and use it for making a quantity of bullets for a machine gun. The gun rotates on a swivel and shoots out the bullets in every direction at random, so if you stand right next to the machine gun you will be hit for sure. However, even if you stand 100 yards away you are not completely safe. Unlike the shock-wave explosion, there is a small but finite chance that a bullet will be lethal. We can imagine assigning a mathematical probability that a bullet will be found far from the gun—a probability distribution over space similar to the probability waves of the quantum theory. Although this probability diminishes with distance from the gun, a bullet either hits or does not, just as an electron is either found or not.

Max Born's machine gun illustrates how probabilities distributed over space and time are not foreign even to classical physics. Probabilities in classical physics possess an extremely important feature—they are linearly additive for independent events. For example, suppose a person is in a house with a front door and a back door. If the probability that the person leaves the front door in the morning and goes to the market is p_1 and the probability for the independent event that he leaves by the back door and goes to the market is p_2, then the total probability that he will go to the market in the morning is $p = p_1 + p_2$—the probabilities simply add. This simple addition of probabilities—almost self-evident—is not valid in the quantum theory. Probabilities in quantum theory have no classical analogue, because they are simply not linearly additive; they are nonlinear.

To understand the nonlinear nature of probability in the quantum theory we will have to go further into Born's statistical interpretation of the de Broglie–Schrödinger wave function. Born's view was that electrons were particles—distinct entities—but their behavior was described by a probability wave. Quantum theory determined precisely the shape and movement of the probability wave.

Look at the waves of the ocean and you see that individual waves seem to move right through each other. The amplitudes or the heights of the waves simply add up to give the total height. If one wave is at a crest at one point and another is at a trough at the same point, the two waves add to zero. This principle of adding individual wave amplitudes to get the total is called the superposition principle—and it lies at the heart of quantum theory.

Like ordinary waves, the probability waves in the quantum the-

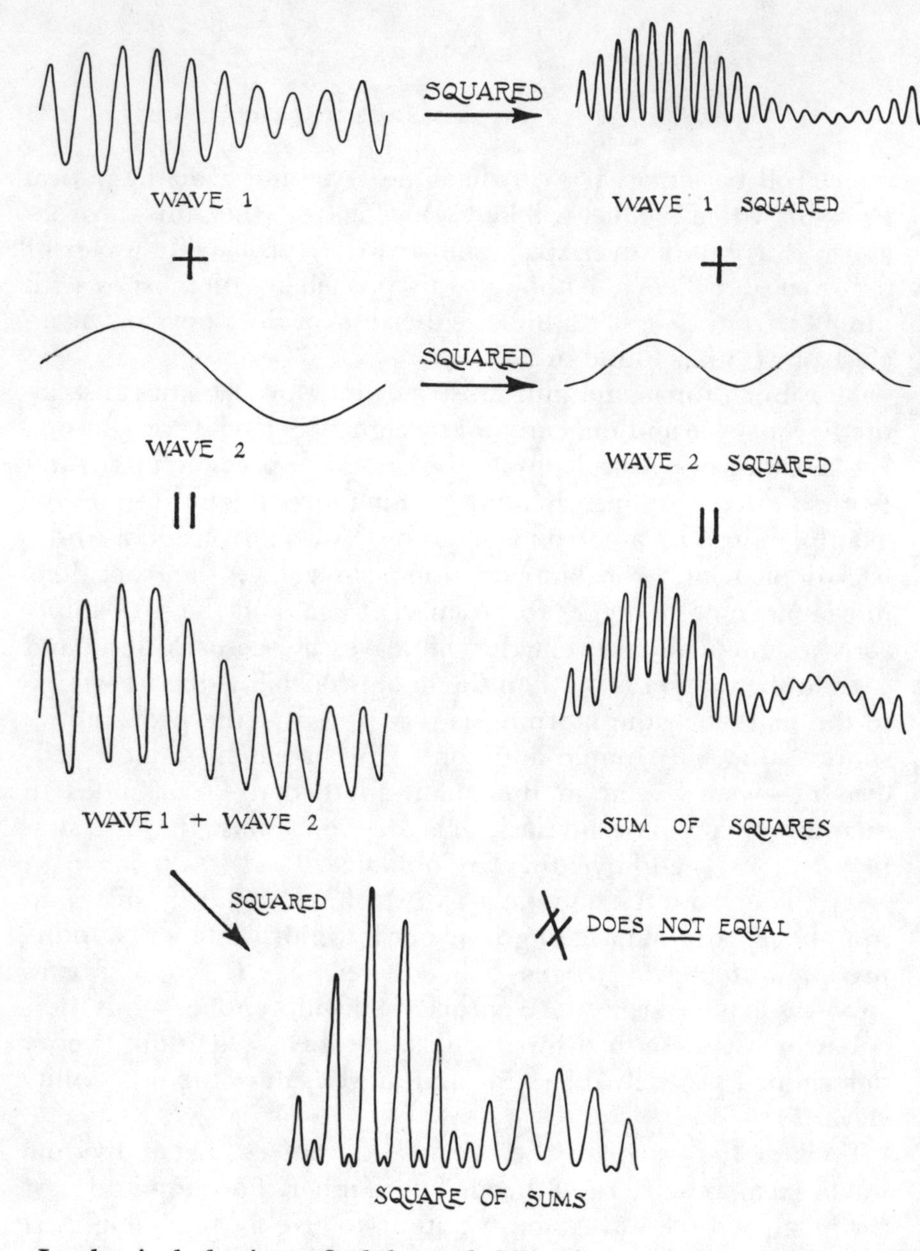

In classical physics to find the probability for two independent events one simply adds the probability for each event. But according to the quantum theory and the superposition principle, one must first add the wave amplitudes for independent events—like an electron going through different holes—and then take the square to obtain the total probability. As this illustration shows, the result is quite different from what one expects from classical physics: The square of the sum is not equal to the sum of the squares. This is why quantum particles do not obey the usual, classical laws of physics and instead exhibit quantum weirdness.

ory obey the superposition principle—if there are two probability waves in some region of space the amplitudes of these waves simply add to give the total height. But there was a further aspect to Born's interpretation. The probability for finding a particle at a point in space is *not* given by the height of the wave at that point but rather by the intensity of the wave—the height of the wave squared, gotten by multiplying the height at that point times itself. If you multiply any positive or negative number times itself, the result is always a positive number. Hence the intensity of a wave is always a positive quantity, and Born identified this with the probability for finding a particle—also always a positive quantity. Because the wave amplitudes add by the superposition principle but the probability for finding a particle is given by the square of the wave amplitude—its intensity—probabilities in quantum theory are not additive, and there results a nonlinear aspect to quantum probabilities, unlike those encountered in classical physics.

For example, suppose we have an electron in a box—like the person in the house—and on top of the box are two holes near each other—like the doors in the house. If you close one hole the probability the electron gets out of the box and is detected at some point outside is p_1 and if you close the other hole the probability of finding the electron at the same point is p_2. But if you open both holes the probability of detecting the electron at this point outside the box, p, is not $p_1 + p_2$ as it is for the person leaving the house. The reason for this peculiar behavior is that the superposition principle implies that the waves of probability associated with the electron coming from each hole can interfere either constructively or destructively with each other, and hence the total intensity that gives the total probability can be either greater or less than that of each wave individually.

This interference of probability waves has no analogue in the classical world of everyday sense awareness and is the fundamental basis for quantum weirdness. The most lucid illustration of this aspect of quantum weirdness is the two-slit experiment. If we examine this experiment closely we can see how Born's statistical interpretation and the superposition principle taken together imply an observer-created reality.

Back in the 1950s, the physicist Richard Feynman gave a series of popular lectures sponsored by the BBC. In one of these lectures he described the two-slit experiment, which consists of a source located some distance from a barrier behind which sits a detection screen. He considered a set of three experiments using machine-gun bullets, waves, and electrons.

First imagine that the source consists of Max Born's machine gun firing bullets at the barrier, a piece of armor plate with two small holes or slits. Behind the barrier stands a sheet of thick wood as a detection screen to catch the bullets. Label the holes 1 and 2. We close hole 2 at the start of the experiment, and bullets are fired at the armor plate. Some bullets get through hole 1, and we measure their distribution as they hit the wood screen. Call this particle distribution P_1. The same is done by closing hole 1 and opening hole 2, and we find a similar distribution called P_2. Next we open both holes. The resulting distribution of bullets hitting the wooden screen is P, and we see that it is simply given by the sum $P = P_1 + P_2$. The bullets as particles are simply obeying the laws of classical physics, and the total probability distribution for hitting the screen is the simple sum from each hole.

In the next experiment we set up the same configuration, but instead of bullets we use water waves. The source now consists of a paddle wheel in water which generates a nice wave that hits the barrier with the two holes. For the detector of the water waves behind the barrier, we can imagine using a screen made of lots of cork bobs. Counting how often the cork bobs jump determines the intensity of the wave at the screen. If we close hole 2 and measure the wave intensity at the screen we obtain the wave intensity distribution W_1, which looks like P_1, which we got from the bullets. Similarly we obtain W_2, which is like P_2. But now if we open both holes 1 and 2 the resulting distribution W looks nothing like P gotten from the machine gun—it is not the simple sum $W_1 + W_2$. Instead it has all sorts of wiggles which are a result of the interference of the waves from holes 1 and 2 with each other.

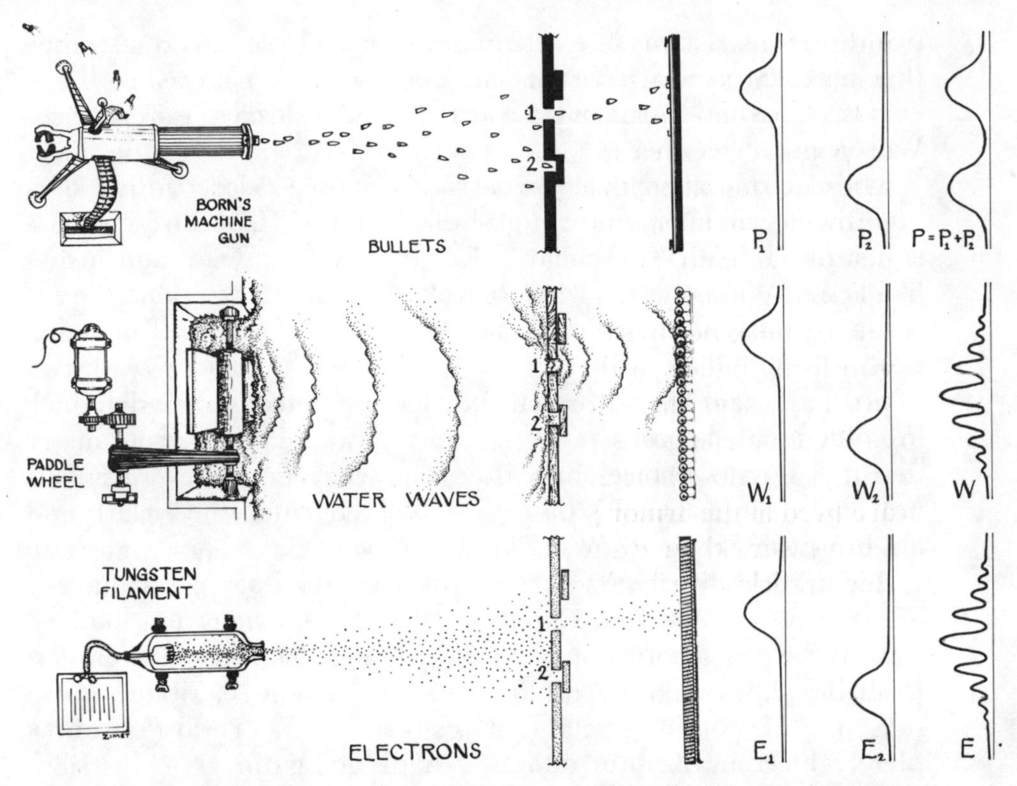

The three different two-hole experiments described in the text: Max Born's machine gun, water waves and electrons. While we can visualize how the bullets and the water waves (both classical physics objects) produce the observed patterns at the detecting screen, we cannot visualize what happens to electrons (quantum particles) at the two holes in order to produce their observed pattern at the screen.

The waves from holes 1 and 2 can cancel or enhance each other. The difference between the two results is that waves can interfere with each other but bullets cannot.

So far all our experiments have been classical physics experiments. Nothing weird here—everything happens before your eyes.

In the next experiment we use electrons as our projectiles, and the source is a hot tungsten filament that boils off electrons, the barrier is a sheet of thin metal with two holes, and the screen is a two-dimensional array of electron detectors. There is no question that an electron is a true particle, because we can measure the charge, mass, and spin of an electron and it leaves tracks in a Wilson cloud chamber.

The electrons are shot at the barrier with hole 2 closed, and like little bullets, all must go through hole 1, so their distribution, E_1, is just like P_1 and W_1. Similarly, E_2 can be determined, and it is like P_2 and W_2. However, when both holes 1 and 2 are opened the resulting distribution E looks like W—the water-wave distribution!

It is important to realize that the electron detectors are detecting individual electrons at the screen position. The electrons are detected as true particles, and after many detections the distribution looks like E. On this basis we would conclude the electrons are behaving like waves. What kind of wave?

According to quantum theory, the electron does not behave like a water wave or like a matter wave. What *is* behaving like a wave is the probability amplitude for finding electrons—waves of probability! Although the electron is a particle, the distribution of these particles on the screen imitates that of a water wave—completely different from the bullets. What is going on?

Let us first examine this logically and consider the proposition: "The electron goes either through hole 1 or through hole 2." If we measure the properties of a single electron it is indeed a particle, and, if we further believe in classical objectivity, then the above proposition must be true, just as in the case of the machine-gun bullets which must go through either hole 1 or hole 2. Since the electron is a particle, if the world is objective the electron always remains a particle and clearly has to go through hole 1 or hole 2. But if the above proposition is true, then we won't get the distribution we experimentally see, the wave distribution; we must get the particle distribution pattern. We therefore conclude that for the real world the proposition is either false or meaningless.

The notion of classical objectivity is at stake here—quantum weirdness has entered our simple experiment. According to the Copenhagen interpretation the above proposition that the electron goes either through hole 1 or hole 2 is meaningless until we actually set up a detection system next to the holes to see which hole it goes through. You cannot talk about events in the world without actually observing them—the super-realism of Niels Bohr. This is completely different from the classical world view that presumes the objectivity of the world even if we do not observe it: A particle remains a particle and *must* go through either one of the holes.

What actually happens to the electrons as they approach the barrier cannot be visualized. If you try to visualize what happens to an electron as it approaches the holes and try to figure out what happens you will, as Feynman remarked, go "down the drain." If we try to visualize the electron as a little bullet, then we should get the bullet pattern. But we don't. If we try to imagine the electron as some kind of wave, then we should detect waves on the screen. But we don't—we detect individual particles. Visually there is a paradox, because we are trying to fit a picture of objectivity that is in our heads—a fantasy—with the real world. The Copenhagen interpretation, especially as formulated by Bohr, insists that such fantasies are meaningless because they do not correspond to anything which can be implemented in the real world. To find out what the world of quantum reality is like you cannot fantasize but must specify exactly how you observe it—get down to "nuts and bolts." Let us become super-realists and examine directly what happens at the holes.

We set up little beams of light right behind the two holes. Now we can look and see which hole the electron goes through by detecting the light scattered by an electron as it emerges from a hole. However, the moment we turn on the little light beams the original experimental conditions have been altered and the distribution of electrons on the screen changes—the slippery property of quantum reality prevents a paradox. If we know exactly which hole each electron passes through, then the distribution on the screen becomes just like that of the machine-gun bullets—a particle distribution. Once we decide to experimentally check that indeed the electron is a particle going through the hole it behaves like a particle going through the hole.

Suppose we dim the light beams so that we see only a few

electrons as they pass through a given hole. Then the particle distribution begins to shift over to the wave interference pattern in a continuous way. Our little experiment with light beams demonstrates what the observer-created reality means. Whatever we do, we cannot know which hole the electron passes through and still retain the wavelike distribution pattern. Mutually exclusive experimental arrangements—you can have the light beam or not —give mutually exclusive results—the electron behaves like a particle or it does not.

I would like to emphasize that what is at issue here is the nature of physical reality. There is no meaning to the objective existence of an electron at some point in space, for example at one of the two holes, independent of actual observation. The electron seems to spring into existence as a real object only when we observe it! We cannot sensibly speak about which hole it goes through unless we set up an apparatus to actually detect it. Quantum reality is rational but not visualizable.

The two-hole experiment is described by the quantum theory in terms of Born's waves of probability, and the electron is described by such a wave. When the probability wave hits the barrier, part of it goes through hole 1 and part through hole 2—just like the water wave. The wave is actually at both holes—no single particle can do that! The two waves that emerge from the holes obey the superposition principle; they just add up and produce the interference pattern at the detecting screen. This intensity pattern is the probability distribution for detecting individual electrons at the screen. The quantum weirdness lies in the realization that as long as you are not actually detecting an electron, its behavior is that of a wave of probability. The moment you look at the electron it is a particle. But as soon as you are not looking it behaves like a wave again. That is rather weird, and no ordinary idea of objectivity can accommodate it.

Our analysis of the simple two-hole experiment shows how Born's statistical interpretation of the quantum waves not only means the end of determinism but the end of classical objectivity. The ancient idea that the world actually exists in a definite state has come to an end. The quantum theory reveals a new message —reality is in part created by the observer. This new aspect of reality vindicates the scientific faith that human reason can understand the world even when human senses must necessarily fail to grasp it.

While the two-hole experiment is simply a thought experiment which dramatically illustrates quantum weirdness, there are many practical devices that actually use the weird properties of waves of probability. A practical example of quantum weirdness is the quantum mechanical tunneling phenomenon—the tunneling or transport of atomic particles like electrons from one side of a barrier to the other, right through a wall. This effect of particles materializing where they cannot be according to the laws of classical physics is understood only by use of quantum theory.

Imagine an ordinary particle sitting inside a barrier like a marble in an empty cup. If there is no force applied to the marble, it

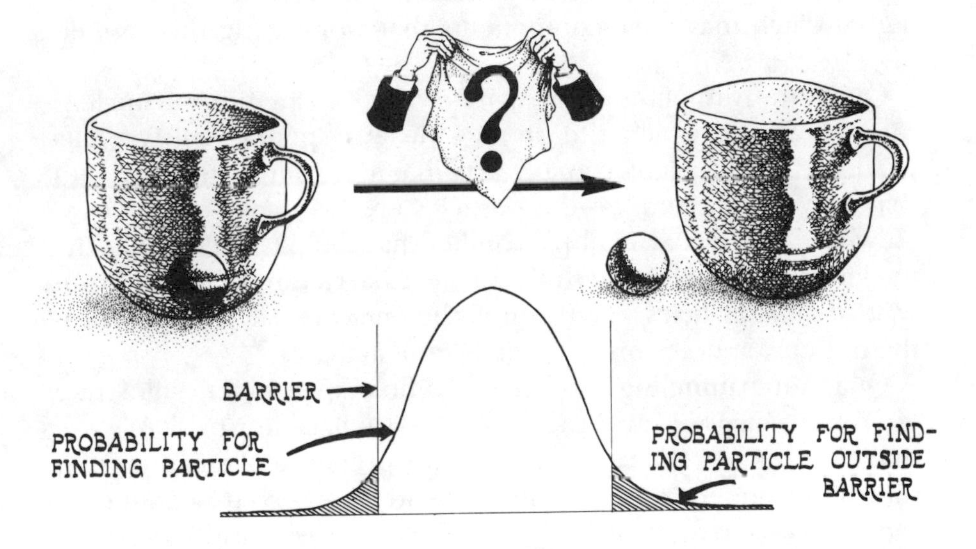

Quantum mechanical tunneling through a barrier—another illustration of quantum weirdness. We cannot visualize how it happens, yet there is a definite probability that a quantum particle can pass through a wall, like a marble passing through the wall of a cup.

cannot escape from the cup—it is trapped. But in the quantum theory we must describe the particle by a probability wave inside the cup, and the intensity of this wave gives the probability of finding the particle. Suppose the particle is an electron. Remarkably, if you solve Schrödinger's wave equation for the electronic wave shape inside the barrier its solution has a little bit of the wave leaking outside of the barrier. This implies that the electron has a certain probability of appearing outside of the barrier—like stepping through a wall—and this is called quantum mechanical tunneling through the barrier. We cannot visualize what happens, but nonetheless particles really do materialize on the other side of the barrier where they are not allowed to be by the old classical physics. Electronic engineers have found that the quantum tunneling effect can be used to amplify electronic signals, a feature used in many practical devices. Transistors, tunnel diodes, and many electronic instruments utilize the properties of waves of probability and the quantum weirdness of electrons—even your digital watch may have components that work using this tunneling effect.

Quantum tunneling is also part of the explanation of nuclear radioactivity in which the atomic nucleus spontaneously emits particles. The nucleus actually acts like a barrier to the particles that it eventually emits—the particles are like marbles in a cup. However, there is a small probability that the particles inside the nucleus can tunnel through the nuclear barrier and escape. From time to time particles step through the nuclear wall, fly away from the nucleus, and are manifested as radioactivity.

Quantum tunneling and the two-hole experiment illustrate quantum weirdness and the end of a visualizable world. We see that according to the Copenhagen interpretation of the quantum theory the indeterminate universe had another consequence—the observer-created reality. The idea that the world exists in a definite state independent of human intention came to an end. There is something out there in the quantum world; we can tame it with our mathematics. But the quantum world is certainly weird—weirder than we can visually imagine.

We are back on the road to quantum reality to continue exploring. Most of the physicists in our group of pilgrims are very comfortable with the Copenhagen interpretation. But one of them, a lean, bespectacled Austrian, carries a large, well-sealed box. Someone asks him, "What's in the box?" He only responds with a quizzical look that indicates, "Wouldn't you like to know too?" We sense that we are about to learn something new about quantum reality.

10. Schrödinger's Cat

"You'll see me there," said the Cat, and vanished.
Alice was not much surprised at this, she was getting
so well used to queer things happening.
—LEWIS CARROLL, *Alice in Wonderland*

IN THE 1980s a new generation of high-speed computers will appear with switching devices in the electronic components which are so small they are approaching the molecular microworld in size. Old computers were subject to "hard errors"—a malfunction of a part, like a circuit burning out or a broken wire, which had to be replaced before the computer could work properly. But the new computers are subject to a qualitatively different kind of malfunction called "soft errors" in which a tiny switch fails during only one operation—the next time it works fine again. Engineers cannot repair computers for this kind of malfunction because nothing is actually broken.

What causes the soft errors? They occur because a moderately high-energy quantum particle may fly through one of the microscopic switches, causing it to malfunction—the computer switches are so tiny they are influenced by these particles which don't disturb larger electronic components. The source of these quantum particles is the natural radioactivity in the material out of which the microchips are made or cosmic rays raining down on earth. The soft errors are part of the indeterminate universe; their location and effect are completely random. Could the God that plays dice trigger a nuclear holocaust by a random error in a military computer? By shielding the new computers and reducing their natural radioactivity the probability for such an event can be made extremely small. But this example raises the question of whether the quantum weirdness of the microscopic world can creep into our macroscopic world and influence us. Can quantum indeterminacy affect our lives?

The answer is yes—as the example of the soft errors in computers shows. Another example is the random combining of DNA molecules at the moment of conception of a child, in which quantum features of the chemical bond play a role. Atomic events which are completely unpredictable deeply influence our lives—we are in the hands of the God that plays dice.

Unquestionably, quantum indeterminacy can influence our lives. But now a puzzle arises if we think about the implications of the two-hole experiment. The standard Copenhagen interpretation of this experiment showed that indeterminacy—Born's waves of probability—meant that we had to renounce the objectivity of the world, the idea that the world exists independent of our observing it. For example, the electron exists as a real particle at a point in space only if we observe it directly. The puzzle is that if indeterminacy implies nonobjectivity and if the macroscopic human world is influenced by indeterminant events, does this mean that human-scale events lack objectivity—that they exist only if we observe them directly? Do we have to renounce the objectivity not only of an electron passing through a hole but also of the annihilation of the entire human species?

Remarkably, if we adhere strictly to the Copenhagen interpretation of the quantum theory, then the quantum world's weirdness can creep out into everyday reality—the whole world, not just the atomic world, loses objectivity. Erwin Schrödinger devised a clever thought experiment he called the cat-in-the-box to show how crazy the Copenhagen interpretation really was and that it required the whole world to possess quantum weirdness. Unfortunately his intent in this experiment, which was to criticize the Copenhagen interpretation, has been more often misunderstood than understood. Some people who want to see the weird reality of the quanta manifested in the ordinary world have used Schrödinger's experiment to show that this must be so. But they are mistaken. Mathematical physicists have carefully analyzed the cat-in-the-box experiment, especially the physical nature of observation, and arrived at the conclusion that although the macroworld is indeterminate it need not be nonobjective, unlike the microworld. To understand how this is possible, we will first describe a version of Schrödinger's cat-in-the-box experiment and see how indeed it seems to imply the end of the ordinary world's objectivity. Then we analyze more closely the physical act of observation and arrive at the alternate view that we need not apply the Copenhagen interpretation to the macroscopic world—quantum weirdness is only in the microworld.

Schrödinger suggested that we imagine that a cat is sealed in a box along with a weak radioactive source and a detector of radioactive particles. The detector is turned on only once for one minute; let us suppose that the probability that the radioactive source will emit a detectable particle during this minute is one out of two = 1/2. Quantum theory does not predict the detection of this radioactive event; it only gives the probability as 1/2. If a particle is detected, a poison gas is released in the box and kills the cat. The well-sealed box is far away on an earth satellite, so we don't know if the cat is alive or dead.

According to the strict Copenhagen interpretation, even after the crucial minute has passed we cannot speak of the cat as in a definite state—alive or dead—because as earthbound people we have not actually observed if the cat is alive or dead. A way of describing the situation is to assign a probability wave to the physical state of a dead cat and another probability wave to the physical state of the live cat. The cat-in-the-box is then correctly described as a wave superposition state consisting of an equal measure of the wave for the live cat and the wave for the dead cat. This superposition state for the cat in the box is characterized not by actualities but by probabilities—macroscopic quantum weirdness! It is as meaningless to talk about the cat's being alive or dead as it is to talk about which hole the electrons go through in the two-hole experiment. The statement "The electron goes either through hole 1 or hole 2" is also meaningless. The electron, if you do not observe which hole it goes through, exists in a superposition state of equal measure of a probability wave for going through hole 1 and through hole 2. Maybe you can accept that weirdness for electrons. But here we have the same kind of statement, "The cat either is dead or the cat is alive," for a cat, not an electron. Cats, like electrons, can be in a quantum never-never land.

Now let us suppose that a space shuttle with a group of scientists goes out to examine the contents of the orbiting cat-in-the-box and when they open the box they are greeted with a loud meow—the cat is alive. The Copenhagen interpretation of this event is that the scientists by opening the box and performing an observation have now put the cat into a definite quantum state—the live cat. This is analogous to examining with light beams the location of the electron at hole 1 or hole 2. For the scientists in the space shuttle, the state of the cat is no longer a superposition of waves for live and dead cat. But because their telecommunications system has broken down the scientists back on earth don't yet know if the cat is alive or dead. For these earthbound scientists, the cat-in-the-box plus the scientists on board the space shuttle who now know the state of the cat are all still in a probability wave superposition state of live cat and dead cat. The quantum never-never land of the superposition state is getting bigger.

Finally, the scientists on board the space shuttle manage to open a communication link to a computer down on earth. They communicate the information that the cat is alive to the computer, and this is stored in its magnetic memory. After the computer receives the information but before its memory is read by the earthbound scientists, the computer is part of the superposition state for the earthbound scientists. Finally in reading the computer output the earthbound scientists reduce the superposition state to a definite one. Then they tell their friends in the next room, and so on. Reality springs into being only when we observe it. Otherwise it exists in a superposition state like the electron going through the holes. Even the reality of the macroscopic world does not have objectivity until we observe it according to this scenario.

Weird as it seems, this is the standard Copenhagen interpretation of reality. We see that it requires a definite line between the observed and the observer, a split between object and mind. At first this line was between the cat-in-the-box and the space-shuttle scientists. After they opened the box the line moved to between the spacc-shuttle scientists and the computer, and so on. As information about the state of the cat propagated from place to place, so did the objective reality of the live cat. The Copenhagen interpretation demands that a distinction be made between the observer and observed; it does not say where the line between them is drawn, only that it must be drawn.

Something unsettles us in this account of the cat-in-the-box experiment. Somehow we may feel that the microworld of atoms lacks standard objectivity. But should this weirdness get out into the ordinary world of tables, chairs, and cats? Do they exist in a definite state only if we observe them as the Copenhagen interpretation would have it? The analysis of the cat-in-the-box experiment suggests that an observation requires consciousness. Some physicists are of the opinion that the Copenhagen view actually implies that consciousness must exist—the idea of material reality without consciousness is unthinkable. But if we examine closely what an observation is, we find that this extreme view of reality—that tables, chairs, and cats lack definite existence until observed by a consciousness—need not be maintained. The Copenhagen view, while necessary for the atomic world, does not have to be applied to the world of ordinary objects. Those who do apply it to the macroworld do so gratuitously. Let us now examine what actually happens when we observe.

If we observe something, our eyes are receiving energy from that object. But the important feature of an observation is that we obtain information—we know something about the world we didn't know before the observation. In our study of statistical mechanics we learned it is not possible to obtain information without increasing entropy—the measure of disorganization of physical systems. The price we pay for obtaining information is scrambling up the world somewhere else, thus increasing entropy

—an inevitable consequence of the second law of thermodynamics. This increase of entropy implies that time has an arrow—there is temporal irreversibility and physical processes exist which can store information; memory is possible. We conclude that irreversibility in time is the principal feature of observation, not consciousness of the observation, although that, of course, also entails irreversibility because it involves memory. Observations can be carried out by dumb machines or computers, provided they have some primitive memory storage. The main point in this analysis of observation is that once information about the quantum world is irreversibly in the macroscopic world, we can safely attribute objective significance to it—it can't slip back into the quantum never-never land.

In the cat-in-the-box experiment, the information is part of the macroscopic world once the cat is dead or live irrespective of whether or not you actually observe the cat. You can't erase that information, because death is irreversible. For the two-hole experiment, by contrast, the information as to which hole the electrons pass through becomes part of the macroworld only if we set up light beams for observation. The electron, in contrast to the cat, cannot carry any record or memory of what state it is in—which hole it goes through.

Recall the line we drew in the zoom-lens sequence of a smoking pipe between the microworld of smoke particles and the macroworld of recognizable objects. The irreversibility of time came about because we sacrificed specific information about individual particles in favor of relevant averages. This is also what we are doing when an observation is performed such as with the light beams at the two holes. The detailed knowledge of the individual probability waves that describe the electron is being reduced to a specific one. The line between the macroworld and microworld is the same as the line between the observer and the observed. By examining where an irreversible interaction corresponding to an observation has been made we can in most cases draw the line between quantum weirdness and the macroscopic world quite close to atomic phenomena. We conclude that while it is consistent to talk about the quantum weirdness of waves of probability superimposing in the macroworld as we did in the description of the Schrödinger's cat experiment, we are not compelled to do so.

The two-hole experiment and Schrödinger's cat-in-the-box are thought experiments—way stations on the road to quantum reality along which we are traveling. We have learned that quantum theory implies an indeterminate universe not only on the atomic level but also on the level of human events. The Copenhagen interpretation of the two-hole experiment then goes on to imply that we must renounce classical objectivity for quantum particles. If we apply this same interpretation to the cat-in-the-box experiment it seems we must also renounce the objectivity of our familiar world of tables and chairs. But that is carrying the Copenhagen interpretation too far. From our study of the second law of thermodynamics we saw that the difference between the microworld and macroworld was not simply quantitative—a difference in size—but qualitative—time's irreversible arrow which is evident in the macroworld does not exist in the microworld. We find, in fact, the irreversibility of observation means that the world of electrons and atoms is qualitatively different from the world of tables and chairs. Quantum weirdness does not exist for the macroworld. We must superimpose probability waves for an electron to go through holes 1 and 2, but we need not superimpose live and dead cats.

As we continue along the road to quantum reality, we see other way stations along the road—traveler's inns where alternatives to the Copenhagen interpretation of quantum weirdness are offered as food for thought. After finishing lunch in one inn we meet a storyteller who begins his story—a quantum mechanical fairy tale.

11. A Quantum Mechanical Fairy Tale

Let him [the child] know his fairy tale accurately, and have perfect joy or awe in the conception of it as if it were real; thus he will always be exercising his power of grasping realities. . . .

—John Ruskin, introduction to *German Popular Stories,* 1868

Once upon a time far, far away there lived a beautiful princess in a castle high upon a mountain. Her father, the king of a vast realm, had decided that it was time for his daughter to marry, and the question arose as to who should have her hand. To settle this question the king consulted with his adviser, the court magician, who recommended a contest among the suitors.

The contest was a game of skill and chance, which appealed to the king's imagination. The magician produced three small boxes and said that the boxes contained two identical white pearls and a black pearl, one pearl in each box. He opened two of the boxes and displayed a white pearl and a black one. Closing these boxes, he then opened the third box, and it contained the other white pearl.

The task of a suitor was to guess the contents of all three boxes. He would have to point to each of the three closed boxes in any order and say, "This box contains a white pearl, this other box a white pearl, and this last box a black pearl." The magician would open the boxes and reveal the contents. If the suitor guessed correctly he would win the hand of the princess. However, if he failed to guess correctly he would be beheaded—conditions designed to discourage all but the most serious suitors.

Since the princess was extremely beautiful and her father was very rich, many young men came to meet the challenge of the contest. Invariably the magician would first open the two boxes the suitor said contained the two white pearls. But alas, one pearl would be white and the other black. Many a suitor lost his head.

As the pattern became clear, few young men came forward for the contest. Years passed, and the princess was still unmarried. One would have logically supposed that one suitor out of many would have guessed correctly. But this was not the case. Like most people, the suitors all believed in the objectivity of the physical world; they believed the fact that they were choosing—performing a measurement—should not influence the contents of the boxes.

Finally there came a handsome youth from a far part of the kingdom. In this part of the kingdom the canonical educational system of classical logic and measurement theory had not penetrated. The young man was presented with the three closed boxes. Pointing to two of the boxes, he said, "These boxes contain a white pearl and a black one. I will not tell you what the third box contains." Before anyone could object to this breach of the strict rules of the contest, the princess, who by now was impatient with the contest and had taken a rather strong liking to this young man, rushed up and opened the two boxes. Out rolled a white pearl and a black one just as he had said. She then ordered the magician to open the third box, but this box would not open.

The king, a wise man, cast the evil magician out of the court for his trickery. The young man and the princess were married that very day and lived happily ever after.

This quantum mechanical fairy tale is the invention of Ernst Specker, who, together with Simon Kochen, worked on the logic of the quantum theory. In 1965–1967 they published a series of papers which demonstrated the impossibility of employing both classical logic and classical measurement theory in the quantum theory. Classical logic, which is manifested in the grammar of everyday language, is formulated by mathematicians as Boolean logic—the laws of thought. Classical measurement theory assumes the objectivity of the world even when it is unobserved. If quantum theory is correct, Kochen and Specker showed one had to give up either the application of Boolean logic to the quantum world, or the assumption of objectivity. Put simply, one has to give up either ordinary logic or ordinary reality.

Boxes like those used by the magician in the fairy tale do not exist in the macroscopic world. However, the properties of the magician's boxes precisely imitate those of the quantum mechanical system analyzed by Kochen and Specker—a helium atom in its lowest energy state sitting in a magnetic field. You might want to think of a helium atom as a little spinning top. The pearls in the boxes correspond to different spin components of the helium atom. But, as Kochen and Specker showed, any such classical picture of the helium atom must break down.

We would like to think that the pearls in the boxes, like the spin states of a helium atom, have an objective existence. There is a white pearl in one box, a white one in another, and a black pearl in the third. The usual idea of objectivity is that this state of the boxes has meaning independent of our examining the contents. But like the spin states of the helium atom this is not so. The magician prepared the boxes—you can do this also for the helium atom—so that if the result of the first measurement is a white pearl, the second has to be a black pearl. You could never simultaneously show two white pearls. This is like the electron in the two-hole experiment. You cannot simultaneously observe the electron going through both hole 1 and hole 2. But if you don't look, you can fantasize that it must be going through one of the two holes. Similarly the suitors fantasized that each box contained a pearl which had to be either black or white. If you actually try to verify that each pearl in a box has a definite state of existence, as the impatient princess did, all the boxes will not simultaneously open—the third box will not open unless one of the other two is first closed. Like the spin components of the helium atom, you simply cannot observe all the pearls at once. There is no objective meaning to the contents of all three boxes. The magician could always invalidate a suitor's guess simply by picking the two boxes the suitor thought contained the white pearls—and showing that they actually contained one white pearl, and one black pearl. It was an observer-created reality, created in this case by the magician.

What Kochen and Specker showed was that if a classically minded physicist came along and assumed that observation did not disturb the state of an object—that the objective world exists independent of our observation—and if he carried out a series of observations on the spin states of a helium atom, he would, using straightforward Boolean logic, eventually be forced into contradictory statements such as "The helium atom is in the state *A* and the helium atom is not in the state *A*." Once one accepts a logical contradiction one can prove anything one likes—it is the end of rational thought.

There are two ways to avoid such a contradiction. One way is to adopt a non-Boolean logic for the statements. Such a "quantum logic" is perfectly consistent, but its rules do not correspond to the "common sense" rules of ordinary Boolean logic. Just as the eye and brain see the world in terms of Euclid's geometry, so, apparently, the human brain thinks in terms of Boolean logic. But just as we can derive perfectly consistent non-Euclidean geometries, we can make consistent non-Boolean logics which change the grammar of "and" and "or." For example, in the two-hole experiment we examined the statement "The electron goes either through hole 1 or through hole 2." In Boolean logic the "either . . . or" in this statement has its usual meaning: either one possibility or the other—they exclude each other. In non-Boolean quantum logic, one abandons this strict meaning of "either . . . or" and adopts a less restrictive meaning. The interesting feature of this alternative is that we can retain the notion that the world exists objectively but we must adopt a new quantum logic in thinking about it.

The second alternative is that we can retain Boolean logic, but give up the idea that the world has a definite objective state independent of whether or not we observe it—the essence of the Copenhagen interpretation. When the tree falls unheard in the forest not only is there no sound, but it is meaningless to talk about the tree having fallen at all. Most physicists adopt this latter alternative: They retain classical, Boolean logic while accepting the weird idea that the world exists only when we observe it. Like most people, physicists do not wish to give up the usual logic of their thinking for another logic. Such mental gymnastics don't seem practical.

The importance of the work of Kochen and Specker is that there is an alternative to such physical weirdness, and that is logical weirdness. But there is no way to avoid weirdness if quantum theory is correct.

As explorers on the road to quantum reality, we thanked the storyteller for his wonderful fairy tale and its provocative moral that our logic is inapplicable to the quantum world and that is why it is so weird. We devoted the whole afternoon in the inn to long discussions on quantum logic and are tired now and decide to spend the night here. After dinner, just as we begin to settle down, the door of the inn bursts open and a man with a wild look on his face rushes in. "I've just been farther down the road, and it's really weird there. Someone told me that Einstein has proved quantum theory implies that telepathy and acausality exist—everything in the universe is connected to everything else. Quantum physicists have vindicated the ancient wisdom of the East!"

"Nonsense!" shouts a member of our group. "You have distorted what Einstein said; Einstein held steadfast to the principle of physical causality. As for comparing ancient Eastern wisdom and modern physics, have you ever asked yourself why quantum theory was discovered in the West out of an equally ancient tradition of theological debate and Talmudic argument? Besides, Buddhism, with its emphasis on the view that the mind-world distinction is an illusion, is really closer to classical, Newtonian physics and not to the quantum theory, for which the observer-observed distinction is crucial." He went on, but his words were lost on the interlocutor, who had already fallen asleep and wasn't listening. The exchange had all of us excited by what we would find the next day farther down the road to quantum reality.

12. Bell's Inequality

It seems hard to look in God's cards. But I cannot for a moment believe that he plays dice and makes use of "telepathic" means (as the current quantum theory alleges He does).

—Albert Einstein

Physicists responded to the new quantum theory in two ways. The first and primary one was the application of the new theory to natural phenomena, which led to the development of the quantum theory of solids, quantum field theory, and nuclear physics. The second direction was more philosophically oriented and focused on the interpretative problems of the new theory.

It is fair to say that the majority of practicing physicists are not very interested in these interpretative problems. Pragmatic theoretical physicists take their motivation from new experiments and ideas related to experiments. They take the Copenhagen interpretation for granted until there is some experimental indication that they should not. The question of interpreting the quantum theory has had little impact on understanding nuclear physics, elementary particle physics, or the construction of transistors and other electronic devices.

In spite of the lack of impact on the practical problems of modern physics, the research done on interpretative questions persists. Physicists and philosophers cannot rid themselves of the question "What is quantum reality?" In asking this question a degree of clarification about the nature of quantum reality has emerged. Over the years a series of thought experiments such as the two-hole experiment and Schrödinger's cat, as well as actual experiments, have been devised that serve the purpose of exposing quantum weirdness—those features of quantum reality that differ from naive realism. Two of these, the EPR experiment and Bell's experiment, have been extensively discussed by physicists and philosophers. They are the basis for our discussion here about the nature of physical reality.

Following Bohr's initial presentation of the Copenhagen interpretation of the quantum theory in 1927, physicists began to recognize the radical nature of the interpretation of reality it proposed. The essence of the Copenhagen interpretation is that the world must be actually observed to be objective. Einstein was among the most prominent critics of this viewpoint. He eventually ceased to criticize the consistency of the interpretation. Instead he focused his attack on the issue of whether or not the quantum theory gave a complete description of reality.

In 1935, Einstein, Podolsky, and Rosen wrote a paper proposing a thought experiment leading to what is often called the EPR paradox. This is a misnomer, since no contradiction is involved; there is no paradox. The EPR article expressed Einstein's view that the standard Copenhagen interpretation of the quantum theory and objective reality are incompatible. He was right. But the main point of the EPR argument was to argue that the quantum theory as it stands is incomplete—there are objective elements of reality that it does not specify. As Einstein later summarized, "I am therefore inclined to believe that the description of [the] quantum mechanism . . . has to be regarded as an incomplete and indirect description of reality, to be replaced at some later date by a more complete and direct one."

Given that no logical flaw exists in the Copenhagen interpretation and no experiment exists that contradicts the predictions of the quantum theory, how did EPR come to this remarkable conclusion? In order to understand their conclusion, we will have to outline the assumptions the three authors made before describing their thought experiment.

We have discussed the assumption of objective reality—that the world exists in a definite state. Bohr in his Copenhagen interpretation and most physicists realized that quantum theory denies this assumption, but Einstein and his collaborators thought that they were too quick in rejecting the idea that at least some measurable properties of the microworld had an objective meaning. They felt that no reasonable idea of reality could completely reject objectivity, and thus objectivity was the first assumption of the EPR team.

Einstein, we know, was distressed by the indeterminism of the quantum theory. But that objection was not the main one that stood in the way of his accepting the theory's picture of reality. A principle of physics that he held even more dear than determinism was the principle of local causality—that distant events cannot instantaneously influence local objects without any mediation. What the EPR argument did without making any assumption about determinism or indeterminism was to show that quantum theory violated local causality. This finding startled most physicists, because they too held the principle of local causality sacred. Let us examine this concept of local causality more closely.

The basic idea of local causality is the following: Events far away cannot directly and instantaneously influence objects here. If a fire breaks out a hundred miles away there is no way it can directly influence you. A second after the fire breaks out a friend may telephone you and tell you about the fire—but that is ordinary causality. Information about the fire has been transmitted by an electromagnetic signal from your friend to you. We can precisely define causality if we imagine constructing an imaginary surface around any object. Then the principle of local causality asserts that whatever influences the object either is attributable to local changes in the state of the object itself or due to energy being transmitted through the surface. That this principle—accepted by all physicists—stands at the center of all our thinking about physics is expressed in Einstein's remarks:

> If one asks what, irrespective of quantum mechanics, is characteristic of the world of ideas of physics, one is first of all struck by the following: The concepts of physics relate to a real outside world. . . . It is further characteristic of these physical objects that they are thought of as arranged in a space-time

> continuum. An essential aspect of this arrangement of things in physics is that they lay claim, at a certain time, to an existence independent of one another, provided these objects "are situated in different parts of space."

What the EPR team did with their definition of objectivity was to point out that quantum theory had either to violate the principle of local causality or be incomplete. Since no one seriously wants to abandon causality, they concluded the quantum theory had to be incomplete. Here is their argument in a nutshell.

Two particles, call them 1 and 2, are sitting near each other with their positions from some common point given by q_1 and q_2 respectively. We assume the particles are moving and that their momenta are p_1 and p_2. Although the Heisenberg uncertainty relation implies that we cannot simultaneously measure p_1 and q_1 or p_2 and q_2 without uncertainty, it does allow us to simultaneously measure the *sum* of the momenta $p = p_1 + p_2$ and the *distance between* the two particles $q = q_1 - q_2$ without any uncertainty. The two particles interact, and then particle 2 flies off to London while 1 remains in New York. These two locations are so far apart that it seems reasonable to suppose that what we do to particle 1 in New York should in no way influence particle 2 in London—the principle of local causality. Since we know that the total momentum is conserved—it is the same before the interaction as after—if we measure the momentum p_1 of the particle in New York, then by subtracting this quantity from the known total momentum p, we deduce exactly the momentum $p_2 = p - p_1$ of particle 2 in London. Likewise by next exactly measuring the position q_1 of the particle in New York we can deduce the position of particle 2 in London by subtracting the known distance between the particles, $q_2 = q_1 - q$. Measuring the position q_1 of the New York particle will disturb our previous measurement of its momentum p_1 but it should not (if we believe in local causality) alter the momentum p_2 we just deduced for the particle far away in London. Hence we have deduced both the momentum p_2 and the position q_2 of the particle in London without any uncertainty.

But the Heisenberg uncertainty principle says it is impossible to determine both the position and momentum of a single particle without uncertainty. By assuming local causality we have done what is impossible according to the quantum theory. It seems as if the quantum theory requires that by measuring particle 1 in New York we have instantaneously influenced particle 2 way off in London. On the basis of this argument EPR concluded that either we admit that quantum theory has such causality violating "spooky action-at-a-distance" or the quantum theory is incomplete and there is indeed a way of simultaneously measuring both position and momentum. Since few physicists like to admit the possibility of such " 'telepathic' means," we should all accept the conclusion that quantum theory is incomplete.

The EPR article caused a great stir among physicists and philosophers. The complacency of the standard interpretation of quantum reality was shaken. No one had previously emphasized these action-at-a-distance effects implied by the usual quantum theory. Were Einstein and his collaborators right in their conclusion that quantum theory could not be the last word on reality? Where is the flaw in their argument? There is no flaw. There is, however, an assumption in the EPR thought experiment that ought to be made explicit. The argument assumes that properties of particle 2 such as its position and momentum in London have objective existence without actually measuring them. The EPR team deduced these properties assuming they had objective significance purely by measurements on particle 1. They then concluded that if quantum theory was right there had to be action-at-a-distance. This conclusion of EPR is sound.

But there is an alternative interpretation of this experiment—the Copenhagen interpretation, which denies the objectivity of the world without actually measuring it. Bohr, who promoted this view, would maintain that the position and momentum of particle 2 have no objective meaning until they are directly measured. If such measurements are carried out they will obey the Heisenberg uncertainty relations in agreement with the quantum theory. Then one avoids the conclusion of action-at-a-distance—instantaneous nonlocal interactions. Einstein, in opposition to Bohr, could never accept the idea of an observer-created reality. Instead he showed that if reality was objective and quantum theory complete, then there had to be nonlocal effects. Since violating causality is so repugnant, Einstein concluded quantum theory was incomplete.

For over thirty years physicists debated the conclusions of the EPR article. Perhaps hiding behind quantum reality lay yet another reality? To attack this question, John Bell, a theoretical physicist at CERN near Geneva, took the next step on the road to quantum reality in 1965. In his paper he did not appeal to the formalism of quantum theory but directly to experiment—he proposed a real experiment, not just a thought experiment. What Bell showed was that the kind of incompleteness of the quantum theory envisioned by EPR was not possible. There were only two physical interpretations of Bell's experiment—either the world was nonobjective and did not exist in a definite state, or it was nonlocal with instantaneous action-at-a-distance. Take your pick of weirdness.

Bell's paper addressed the question of hidden variables—the idea that somehow the usual quantum theory is incomplete and there exists a hypothetical *sub*quantum theory which specifies additional physical information about the state of the world in the form of these new hidden variables. If physicists knew these variables, they could predict the outcome of a particular measurement (not just the probabilities of various outcomes) and even determine the momentum and position of particles simultaneously. Such a subquantum theory would actually restore determinism and objectivity. If we imagine that reality is a deck of cards, all the quantum theory does is predict the probability of various hands dealt. If there were hidden variables it would be like looking into the deck and predicting the individual cards in each hand.

It would seem that if the usual quantum theory is experimentally correct, then it rules out any subquantum theory of hidden variables and a hidden reality. Von Neumann, the mathematician, had a proof that such variables, hiding behind the veil of quantum reality, could not exist, and because of his proof people stopped thinking about hidden variables. Von Neumann's proof was logically flawless, but as Bell first pointed out, one of the assumptions that went into Von Neumann's proof did not apply to quantum theory and therefore the proof was irrelevant. The question of whether quantum theory allowed for hidden variables and causal reality was still open. To this question Bell next turned.

Bell wanted to see what the quantum world would be like if local hidden variables really existed—and here the word "local" is important. Local hidden variables refer to physical quantities which locally determine the state of an object inside an imaginary surface. By contrast, nonlocal hidden variables could be instantaneously changed by events on the other side of the universe. Assuming that any hidden variables are "local" is the assumption of local causality. Using this assumption, Bell derived a mathematical formula, an inequality, which could be checked experimentally. The experiment has been done independently at least half a dozen times, and Bell's inequality—along with its central assumption of local causality—was violated. The world, it seemed, was *not* locally causal! We will subsequently scrutinize this amazing conclusion, but first we will describe Bell's experiment in considerable detail. As someone remarked, "God is in the details," and we find if we look at the details, the experiment reveals a remarkable act of sleight of hand by the God that plays dice.

Bell's inequality applies to a large class of quantum experiments. Before applying this inequality to the quantum world it is useful first to derive the inequality for a purely classical, visualizable experiment. There will be no quantum weirdness in this classical experiment—just as there wasn't any for the machine gun firing at the two holes. The reason for first deriving Bell's inequality for a classical physics experiment is that all the assumptions that go into its derivation can be seen explicitly. There are no "hidden" variables for a classical system—all the cards are laid out on the table.

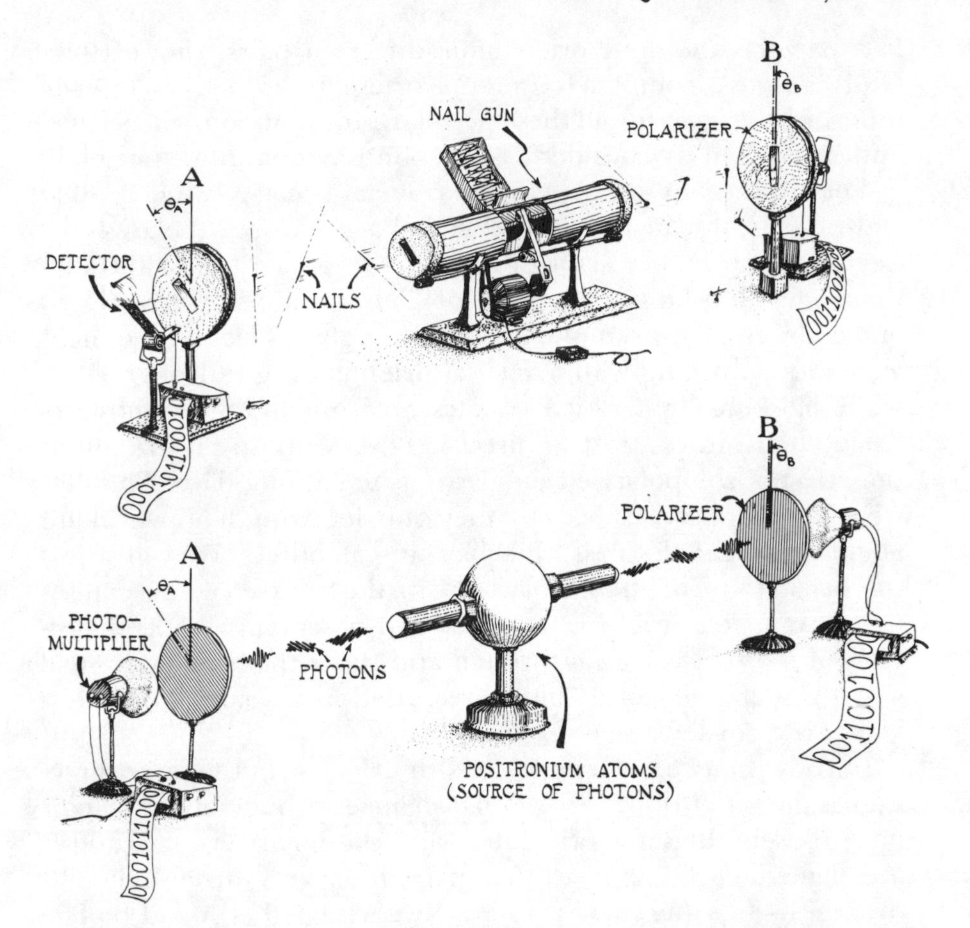

Bell's experiment: The flying nail gun and the positronium source of correlated pairs of photons. If the nails or photons are properly oriented then they pass through their respective polarizers at A and B and are detected. Hits are registered as a 1 and misses as a 0. The angle $\Theta = \Theta_A - \Theta_B$ is the relative angle between the polarizers at A and B.

Imagine that we have a special nail gun that shoots nails two at a time in exactly opposite directions along a fixed line. Unlike most nail guns, which shoot nails like arrows, this one shoots them sideways—a pair of nails flies away from the gun with their long axis perpendicular to the direction of motion. Although each nail in a pair has the same orientation, different pairs, shot off successively, have completely random orientations relative to each other. The reason for all these peculiar requirements will become apparent when we consider a corresponding quantum system.

The flying nails are aimed at two metal sheets, A and B, each with a wide slot in it. These slots behave like real polarizers—devices which let objects with a specific orientation pass through them while blocking the passage of identical but improperly oriented objects. For example, polarized sunglasses let waves of light which are vibrating with a vertical orientation go through them while blocking light which vibrates horizontally. Since most reflected light, in contrast to direct light, is vibrating horizontally, the effect of the polarized sunglasses is to cut out glare. The slots we will call polarizers, because they only let flying nails which are aligned with the slot pass while blocking all others. We can adjust the orientation of these polarizers in the course of the experiment. At sheets A and B there are two observers who keep records of the nails that get through and those that don't. If a nail gets through the slot a "hit" is recorded as a 1 and if it fails a "miss" is recorded as a 0.

Initially the two polarizers are both oriented in the same direction as the gun fires its pairs of nails. Since each member of a pair has precisely the same orientation and the polarizers at A and B are aligned, each member of the pair either gets through the slot or it fails—hits and misses are exactly correlated at A and B. The record at A and B might look like

A: 0100011001000010110100110010110001000100 . . .

B: 0100011001000010110100110010110001000100 . . .

Each sequence of 0s and 1s is random, because the gun fires the pairs out at random orientations. But note that the two random sequences are precisely correlated.

The next step is to change the relative angle between the two polarizers by rotating the slot at A clockwise by a small angle Θ and holding B as a fixed standard. With this configuration a nail in a pair will sometimes get through at A but fail at B and vice versa. The hits and misses at A and B are no longer exactly correlated. The record might look like

```
A: 0001011000101011100011110010110010100100 . . .
     ↕  ↕              ↕     ↕
B: 0011001000101011100011010010010010100100 . . .
```

where the mismatches are indicated. These mismatches we can call "errors," for they may be thought of as errors in A's record relative to B's, which is the standard. In the above example there were 4 errors out of 40, so the error rate E(Θ) for the angle set at Θ is E(Θ) = 10%.

Suppose that we had left the polarizer at A untouched but rotated the one at B counterclockwise by the angle Θ. Now we might say the "errors" are in B's record and A's acts as the standard. The error rate will clearly be the same as before, E(Θ) = 10%, because the configuration is identical.

The final step is to rotate A's polarizer by an angle Θ clockwise so that the total relative angle between the two polarizers is now doubled to 2Θ. What is the error rate for this new configuration? This is easy to answer provided we assume that the errors at A are independent of the situation at B and vice versa. In making this assumption we are assuming local causality. After all, what does a nail getting through its polarizer at A have to do with the situation at B? Since the errors produced at B were previously E(Θ) we must add to this the errors produced by rotating the polarizer at A, which is also E(Θ). So it seems that the error rate with the new setting should be the sum of the two independent error rates, or E(Θ) + E(Θ) = 2E(Θ). But by shifting A by the small angle Θ we have lost the standard record for B's record, and likewise by shifting B we have lost A's standard. This means that from time to time an error will be produced at both A and B —a double error. But a double error is detected as no error at all. For example, suppose a pair of nails would have registered a 1 and 1 at A and B if the polarizers were perfectly aligned. But because A's polarizer is shifted the nail then misses and a 0 is registered. This shows up as an error. But since we have also shifted B's polarizer it is possible that the nail there also misses. This is a double error in which two hits, a 1 and 1, has been changed to two misses, a 0 and 0. The two misses are seen as no error. Because of the impossibility of detecting a double error, the error rate with an angle 2Θ between the two polarizers—$E(2\Theta)$—will necessarily be less than the sum of the error rates for each of the separate shifts. This is expressed mathematically by the formula

$$E(2\Theta) \leq 2E(\Theta)$$

which is Bell's inequality.

No doubt if this odd experiment were performed, Bell's inequality would be satisfied. For example, with an angle of 2Θ the record might look like

A: 0010110011111000101010100111101011101000 . . .
↕ ↕ ↕ ↕ ↕ ↕
B: 0010100011011100101010100110101011001100 . . .

or 6 errors out of 40 so $E(2\Theta) = 15\% \leq 2 \times 10\% = 20\%$. Bell's inequality is satisfied for this classical physics experiment.

Let us examine closely the crucial assumptions that have gone into obtaining Bell's inequality. We have assumed that the nails are real objects flying through space and that the orientation of pairs of nails is the same. We aren't actually observing that the nails have a definite orientation because they fly by us so quickly. This seems like a safe assumption for nails, but we have indulged in the fantasy of objectivity. We are assuming that the nails exist like ordinary rocks, tables, and chairs. Suppose we are the observer at A. Then we are assuming that a nail flying toward B, even if B is on the moon, has a definite orientation. The notion that things exist in a definite state even if we do not observe them is the assumption of objectivity—and of classical physics.

The second crucial assumption in obtaining Bell's inequality was that the errors produced at A and at B were completely independent of each other. By shifting the polarizer at A we did not influence the physical situation at B and vice versa—the assumption of local causality.

These two assumptions—objectivity and local causality—are crucial in obtaining Bell's inequality. What happens if we now replace flying nails with photons—particles of light?

Instead of a nail gun we will use positronium atoms as our source of particles. Positronium is an atom consisting of a single electron bound to a positron (anti-electron), and this atom sometimes decays into two photons traveling in opposite directions. The important features of this positronium decay is that the two photons have their relative polarization precisely correlated—like the flying nails. The polarization of a photon is the orientation of its vibration in space. If one photon has its polarization in one direction, its companion flying off in the opposite direction has its polarization in the same direction. The absolute direction of the polarization of the two correlated photons changes from decay to decay in a random way, but the relative polarization between any pair of photons is fixed. That is the important feature of this source—it is like the nail gun.

The photons fly off in opposite directions and pass through separate polarizers at A and B, located far apart with observers stationed there. Behind the polarizers are photomultiplier tubes that can detect single photons. If a photomultiplier tube detects a photon, the event is recorded by a 1, and if it detects no photon the event is recorded by a 0. In the initial configuration the two polarizers at A and B are perfectly aligned relative to each other. Let the polarizer at B be fixed while the one at A is free to rotate and call the relative angle between the two polarizers Θ so that in this initial configuration $\Theta = 0$.

If a photon hits the polarizer it has a certain probability of getting through and being detected. If the photon's polarization happens to align parallel to that of the polarizer it gets through to the detector, and a 1 is registered. If the polarization of the photon is perpendicular to the polarizer, then it won't get through and a 0 is registered. With other orientations there is only a probability that it gets through.

The polarization of the photons relative to the polarizer is completely random, so that each detector, in the initial configuration with $\Theta = 0$, will register a series of 0s and 1s. Suppose the series looks something like this at each detector:

A: 01101011000010110101110011000101110 . . .

B: 01101011000010110101110011000101110 . . .

This is just like the records with the nail gun. The series are identical because each pair of photons is polarized identically and the angle between the polarizers is zero. Further, each series has on the average an equal number of zeroes and ones, since it is as likely for a photon to get through the polarizer to the detector as not.

Next we rotate the polarizer at A, slightly, so the angle Θ is not zero. Set $\Theta = 25°$. This slight shift means that the two photons in each pair have a slightly different probability of going through the polarizers and being detected. The series are not precisely identical but instead disagree from time to time. However, on the average, both the series at A or B have an equal number of zeroes and ones because the probability of getting through the polarizer is independent of its orientation. The new series looks like

```
A:  0010111011000111110110100111000101011100 . . .
     ↕  ↕           ↕          ↕
B:  0110011011000111010110100110000101011100 . . .
```

where we have indicated the mismatches. In the above example the rate of errors, since there are 4 errors out of 40 detections, is $E(\Theta) = 10\%$.

So far this experiment done with photons resembles that with the nails. Photons are behaving just like the perfectly visualizable experiment with the flying nails. If we assume the state of polarization the photons have at A and B is objective (objectivity assumption) and that what one measures at A does not influence what happens at B (local causality assumption), then Bell's inequality, E(2Θ) ≤ 2E(Θ), ought to hold for this experiment. If we double the angle to 2Θ = 50%, the following records are found:

```
A:  1000111001100110111001111110110101000100 . . .
     ↕↕     ↕  ↕    ↕↕↕        ↕        ↕ ↕↕↕
B:  1110111101000111001001110110110101101010 . . .
```

This is 12 errors out of 40, so that E(2Θ) = 30%. Now let us compare this result with the requirement of Bell's inequality. Since E(Θ) = 10% we have 2E(Θ) = 20%; but Bell's inequality requires that E(2Θ) ≤ 2E(Θ), so that 30% is supposed to be less than 20%—completely false—30% is greater than 20%! We conclude that Bell's inequality is violated by this experiment, as it is for real experiments with photons. Consequently, either the assumption of objectivity or locality or both are wrong. That is very remarkable!

We have described the experiment and Bell's inequality in some detail because it is rather elementary and illustrates the crux of quantum weirdness. Bell was motivated to find a way of testing if there were hidden variables that exist out there in the world of rocks, tables, and chairs. He showed that the violation of the inequality by quantum theory did not necessarily rule out an objective world described by hidden variables but the reality they represented had to be nonlocal. Behind quantum reality there could be another reality described by these hidden variables and in this reality there would be influences that move instantaneously an arbitrary distance without evident mediation. It is possible to believe the quantum world is objective—as Einstein wanted—but then you are forced into accepting nonlocal influences—something Einstein, and most physicists, would never accept.

To get an intuitive sense of how objectivity implies nonlocality, compare the records for the angle Θ = 25° and Θ = 50°. There are just too many errors (12) for the 50° setting as compared to the number of errors (4) for the 25° setting. It seems that by moving A's polarizer we must have influenced the polarization of the photons about to be detected at B and that produces all those "extra" errors that violate Bell's inequality. Observer B could be on the earth and A light-years away, on another galaxy. A, by moving the polarizer, it seems, is sending a signal faster than the speed of light, thus instantaneously changing B's record. That certainly seems like action-at-a-distance and the end of locality.

Now that we see what we have been forced into we might want to look at this a bit further. Either alternative—a nonobjective or a nonlocal reality—is a bit hard to take. Some recent popularizers of Bell's work when confronted with this conclusion have gone on to claim that telepathy is verified or the mystical notion that all parts of the universe are instantaneously interconnected is vindicated. Others assert that this implies communication faster than the speed of light. That is rubbish; the quantum theory and Bell's inequality imply nothing of this kind. Individuals who make such claims have substituted a wish-fulfilling fantasy for understanding. If we closely examine Bell's experiment we will see a bit of sleight of hand by the God that plays dice which rules out actual nonlocal influences. Just as we think we have captured a really weird beast—like acausal influences—it slips out of our grasp. The slippery property of quantum reality is again manifested.

Bohr would be the first to point out an alternative interpretation of the experimental violation of Bell's inequality. In order to conclude that the photons were subject to nonlocal influences we have indulged in the fantasy that they exist in a definite state. Try and verify that, Bohr would insist. If we can verify that the photons actually exist in a definite state of polarization without altering that state, then indeed we must conclude from Bell's experiment that we have real nonlocal influences.

For the flying nails this verification is easy—we set up a high-speed camera and take pictures of them just as they arrive at the polarizers. This won't disturb their state. But then the experiment with the flying nails did not violate Bell's inequality as did the experiment with photons.

If we now try to verify the state of polarization of a photon we find that this is not possible without altering the requirement that both members of a pair of photons have identical polarization. In measuring the polarization of the photon we put it into a definite state, but this alters the initial conditions of the experiment. This is identical to the problem we faced in the two-hole experiment with the electron. By observing with light beams which hole the electron went through we changed the detected pattern. Likewise, the very act of establishing the objective state of the photon alters the conditions under which Bell's inequality was derived. The attempt to experimentally verify the objectivity assumption has the consequence that the conditions of the experiment are altered in just such a way that we can no longer use the violation of Bell's inequality to conclude that nonlocal influences exist.

Suppose then that we do not try to verify the state of the flying photons. After all, we have the records of hits and misses at A and B and these are part of the macroscopic world of tables, chairs, and cats and are certainly objective. Cannot the observer at B read his record, see that Bell's inequality is violated, and conclude that local causality has also been violated? The answer is no, because the God that plays dice has a trick to show us. Remember that the source of photons emits them in pairs with random polarization. This means that the records at A and B, no matter what the angle is, are completely random sequences of 0s and 1s. And that fact is what lets us slip out of the conclusion of real nonlocal influences.

At first you might think that by changing A's polarizer we have directly influenced the number of errors produced at B. Hence by altering A's polarizer to various settings in a sequence of moves, B could, by observing the alteration in the number of errors produced at B, get a message from A—a telegraph that would violate causality. But no information can possibly be transmitted from A to B using this device because holding a single record of events at either A or B would be like holding the message of a top-secret communication in a random code—you can't ever get the message. Because the sequences at A and B are always completely random there is no way to communicate between A and B. That is how real nonlocality is avoided by the God that plays dice; He is always shuffling the deck of nature.

Random stereograms, which we already discussed, illustrate

this trick. Each half of the stereogram is completely random, but two random sequences of dots if compared can yield nonrandom information. The information is in the cross-correlation gotten by comparing the two sequences. It is the same with the records at A and B—the information about the relative angle between the polarizers at A and B is in the cross-correlation of the two records; it is not in either record separately. All that happens when the polarizer angle is changed is for one random sequence to be changed into another random sequence, and there is no way to tell that happens by looking at only one record. Because such real random processes actually occur in nature—as they do in this experiment—we avoid the conclusion of real nonlocality.

What a marvelous trick nature has used to avoid real nonlocal influences! If we asked out of all things in this universe which one, if altered in a random way, would remain unchanged, the answer is: a random sequence. A random sequence changed in a random way remains random—a mess remains a mess. The random sequences at A and B are like that. But by comparing these sequences we can see that there has been a change due to moving the polarizers—the information is in the cross-correlation, not in the individual records. And that cross-correlation is completely predicted by the quantum theory.

We conclude that even if we accept the objectivity of the microworld then Bell's experiment does not imply actual nonlocal influences. It does imply that one can instantaneously change the cross-correlation of two random sequences of events on other sides of the galaxy. But the cross-correlation of two sets of widely separated events is not a local object and the information it may contain cannot be used to violate the principle of local causality.

With Bell's inequality and the EPR experiment we have entered into the essence of quantum weirdness. To see what reality we can buy with these experiments let us go to the reality marketplace.

13. The Reality Marketplace

The true mystery of the world is the visible,
not the invisible.
—Oscar Wilde

We are coming to the end of our road to quantum reality. The road may well go on far into the future of physics, and new insights into the quantum theory may be found. Perhaps quantum theory is experimentally wrong or incomplete, something that is not logically impossible. No doubt there are incredible things yet to be discovered on the road to quantum reality. But in the absence of a new interpretation or the experimental failure of the quantum theory, for us the road has come to a resting place. What we find here is a kind of marketplace—a reality marketplace.

The reality marketplace has lots of shops, each with a merchant who wants to sell us his version of physical reality. The way the market is set up, we have only enough cash to buy one reality, so it is a very competitive market. We are rather sophisticated buyers now, having learned about the two-hole experiment, the EPR and Bell's experiment, the work on quantum logic and Schrödinger's cat. The merchants in the shops know about these too, and nobody disagrees about the actual experiments. It is the *interpretation* of these experiments in terms of physical reality that is being sold. The interpretation of experiment is, however, not decided by experiment. To distinguish realities as buyers, we must invoke other criteria, such as paucity of assumptions, potential empirical content, and taste. There are many shops, but we need only look in a few shops with the finest merchandise. We go into these to listen to the sales pitch.

The first shop we come to, a tent on the outskirts of the marketplace, has the sign "Many Universes for Sale" posted outside, an intriguing come-on. The salesman inside explains there is an easy way out of all these problems posed by the quantum theory. He points out we have a hidden assumption in all our thinking about quantum reality: There is only one reality. "If you are still left wondering," he continues, "which hole the electron 'really' went through or if Schrödinger's cat is dead or alive, then just imagine that the whole universe at every instant of time splits into an infinity of universes. All those quantum never-never lands become real. In some of those universes the electron goes through hole 1 and in others through hole 2. Different universes do not communicate with each other, so no contradiction is possible. Reality is the infinity of all those universes existing in a 'superspace' that includes them all. One can imagine all sorts of incredible things happening in those other universes—everything that can happen, will happen. In some universes none of us were even born to wonder about quantum reality. Stop looking for reasons why our universe exists or has features like life on earth; ours is just that one universe out of an infinity that happens to be the way we find it. Otherwise we wouldn't be here to ask the question."

"But," asks one of our friends, "isn't this multi-reality and superspace all a great fantasy, which although not strictly ruled out by the quantum theory, is not required by it? You speak of all those other worlds as if they were real, when in fact, it is only this world—the one we live in—that we can ever know. And even accepting your many-worlds interpretation of the quantum theory, how do we know that the other worlds are so different from ours? Maybe different worlds are related to each other like the different configurations of molecules in a gas—each configuration is radically different from most of the others, but what matters is that the gas viewed as a whole looks nearly the same for all those different molecular configurations. And didn't John

Wheeler, one of the physicists who helped develop this many-worlds view, finally reject it because, in his words, 'It required too much metaphysical baggage to carry around'? I can't disprove the many-worlds viewpoint," concluded our friend, "and that's just why it doesn't interest me as a physicist."

The salesman looks crestfallen as our group leaves his tent. "Aren't you going to listen to the rest of my sales pitch?" he pleads.

"Perhaps in some other world," says our critical friend, "but not in this one!"

The next shop we come to is the Quantum Logic Shop. The merchant of this shop is going to try to convince us if we buy his "quantum logic" and give up the Boolean logic of ordinary language, then quantum reality will not seem weird. Quantum logic is non-Boolean—the usual logical meaning of words like "and" and "either . . . or" is changed. The merchant in the quantum logic shop tells us the reason the quantum world seems so weird is that we do not think about it correctly—our grammar is wrong and its ordinary logic does not apply to the quantum world. In the shop there are a few demonstration computers and artificial intelligences that have been wired up to think in a non-Boolean, quantum-logic way. Such intelligences find the quantum world "natural" and not at all weird, and this seems very impressive.

The interesting question suggested by quantum logic is how one determines the correct logic for thinking about the physical world. Logic becomes an empirical problem just as previously the geometry of space and time became an empirical problem with the advent of general relativity. We have learned from general relativity that the world is in fact non-Euclidean. The lesson of quantum theory can be interpreted to imply that the logic of the physical world is non-Boolean. Logic, usually thought to be prior to any experience, becomes empirical—depending on our experience—just like geometry.

In spite of the persuasive arguments of the merchant of quantum logic—it seems such a simple solution to all our problems—most physicists, like most people, are reluctant to give up their ordinary Boolean way of thinking. Physicists cannot help thinking this way; it is the way ordinary language corresponds to the world of experience. They feel that adopting non-Boolean quantum logic is a kind of trick—it puts the quantum weirdness in the mind rather than in the physical world where they think it properly belongs.

One of the members of our group tells us the analogy between non-Boolean logic and non-Euclidean geometry is not precise. It is true that according to general relativity the geometry of space is non-Euclidean in a gravitational field. But for weak gravity fields, space becomes very close to ordinary, flat Euclidean space. But logic is not like that—it is an all-or-nothing decision to pick one logic and scrap the other. Once you decide to organize your concept of the physical world according to a logic, that logic, be it Boolean or non-Boolean, must then be applied to the whole world.

On the way out of the Quantum Logic Shop we meet a few people who bought quantum logic. These poor people were so troubled by the weirdness of the quantum world that they got their brains rewired to think in terms of non-Boolean logic. What a mistake they made—they can't even carry out a personal or financial transaction using quantum logic. If we tell them about the two-hole experiment they just smile—they have no idea what the problem is. Now we see what the trouble with quantum logic is—it is more restrictive than ordinary Boolean logic. You cannot prove as much with quantum logic, and that is the reason you do not have any sense of weirdness in the physical world. Adopting quantum logic would be like inventing a new logic to maintain the earth was flat if confronted with the evidence that it is round. The new logic would allow us to interpret the evidence in such a way as to maintain the earth as "flat." Some people actually tried to do that. But it seems that the price of changing our logic is too high.

We notice that the merchant in the Quantum Logic Shop hasn't rewired his brain. He needed ordinary logic to try to convince us that quantum logic was the answer to quantum weirdness. Out we go, back on the road to reality with our minds intact.

Next we come to two shops side by side. They are the most

crowded of the shops in the reality marketplace, the Local Reality Shop and the Objective Reality Shop, which are in direct competition. Some people actually buy from one shop precisely because they know it means renouncing the reality being sold in the other shop—a reality they cannot accept. We decide to go into the Objective Reality Shop just as the merchant is beginning his sales pitch.

"The basis of physics," he begins, "indeed the whole of science is predicated on the existence of an objective reality—a world of objects that exists independent of our knowing it. What is happening on the moon, down the street, or behind your back should not depend on whether or not you observe it. The microworld must also be objective, otherwise how can we pretend to have a science of that world? If you deny the objectivity of the world unless you observe it and are conscious of it, then you end up with solipsism—the belief that your consciousness is the only one. No science can take solipsism seriously. But that is what you must do if you do not accept the objectivity of the microworld.

"Accepting the objectivity of the microworld is not a step backward into classical physics. Einstein knew that. What it means is that quantum theory is incomplete and someday we will discover how to go beyond quantum theory. Maybe beyond quantum reality lie hidden variables that once we know them will restore determinism and we shall then see that our need for statistical methods is only an artifact of our ignorance of a more fundamental theory. New worlds may open up before us once we see how nature has stacked the deck."

While the merchant continues his speech appealing to our reason, some of his less reputable colleagues are working credulous members of the crowd. We overhear one of them in a corner of the shop giving a hard sell to a well-known occultist. "Did you know that if you buy objective reality then you are required to renounce the idea of a locally causal reality? Isn't that exciting? There must then exist—according to the experimental findings of quantum theory—nonlocal instantaneous interactions, just like telepathy."

"I've known all that for a long time," remarks the occultist. "It's amusing to me that materialist scientists have taken so long to see this ancient truth of reality. All of what they call material reality is instantaneously interconnected by nonmaterial forces. Everything in the universe is connected to everything else—reality is nonlocal and acausal. This knowledge is well understood by us who have steadfastly held to the tradition of Hermes Trismegistus, the ancient magician. By their own slow methods, scientists only now discovered the acausal universe. But this is not the time for vindictive self-satisfaction. Quantum physicists and occultists should now join forces to begin the great work of exploring the new reality."

Many people in the crowd are drifting over to the shop corner to take in this dialogue. We realize immediately that the salesman and occultist are collaborators, a team trying to sell a reality to those who want to see the bizarre and incredible realized in science. Many sincerely religious people or those with mystical inclinations have been drawn to these ideas. But it is a setup—the merchant appeals to the scientific skeptics while his colleagues work the nonscientific groups.

Shouting begins in the corner when a salesman from the Local Reality Shop makes the scene. "It is a fraud selling objective reality to those who want to believe in telepathy and an interconnected universe. The quantum theory, the EPR, and Bell's experiment imply nothing of this kind. That's rubbish! Even if you decide to buy objective reality—and I don't recommend it—there is no way that it requires any kind of telepathy or instantaneous long-range interactions. I urge you to use your minds, those of you who haven't bought quantum logic, and listen carefully to my arguments." The occultist and his salesman partner know this fellow all too well and try to shout him down. But the crowd won't let them and begins to listen to the infiltrator's argument.

"Suppose for the sake of our discussion we accept the merchant's view that the microworld reality must be objective. Then it is true that there are nonlocal influences, just as his salesmen say. But what kind of nonlocal influences are these? Consider the analysis we made of Bell's experiment with photons. It seems that moving the polarizer at A instantaneously influences the situation at B so more errors are produced. If we assume the photons are objective and exist in a definite state, then that sure seems like action-at-a-distance and a violation of causality.

"But it really isn't. People who want to believe in such acausality are really closet determinists. In order to restore objectivity and determinism to the world they are willing to hypothesize these nonlocal influences that pervade the universe. This is a modern version of the old ether theory of absolute space held by physicists before Einstein invented special relativity theory. The ether theory maintained that space was a kind of substance that pervaded everything like a transparent jelly. Maybe there is an ether, but we cannot detect it if Einstein is right. Occam's razor—the principle that superfluous assumptions should be eliminated—may be effectively applied. You can forget about quantum theory's implying real acausal influences and you can forget about the ether—they have no experimental consequences."

"Are you telling us that the nonlocal influences implied by our assumption that the photon and other quanta are objectively real cannot be verified if the quantum theory is right?" someone in the crowd asks. A sigh of disappointment is heard from those people who thought that quantum theory vindicated telepathic communication around the universe. The occultist has left the shop in disgust. The new speaker is very persuasive and has the crowd charmed.

"That is right," he continues. "It is true that in the quantum theory, it *appears* as if there is instantaneous action at a distance, and this is what so distressed Einstein. Changing the polarizer at A in Bell's experiment instantaneously influences the record at B. But I repeat my question, 'What kind of an influence is this?' Although the record at B is influenced by rotating A's polarizer, there is no way to tell this by just looking at the record at B! The reason is that the source emitting the photons emits them with random polarization. Therefore the record at A and B, the sequence of 0s and 1s, is *completely random* no matter what the angle

Θ between A's and B's polarizer. A single random sequence at A or B contains no information whatsoever. A single sequence is like the coded message of a top-secret communication without the decoding key sequence. However, the cross-correlation between the sequences at A and B is *not* random and depends on the polarization angle, Θ. The remarkable fact is that two completely random sequences such as those at A and B when compared can yield nonrandom information.

"Because each pattern is truly random we cannot use Bell's experiment to uncover real nonlocality—only a kind of after-the-fact nonlocality. And the latter only if we accept the objective existence of the photons irrespective of whether we actually observe their state. It is because of the randomness of each record that real nonlocal influence or action-at-a-distance is impossible. The God that plays dice has intervened to prevent acausality. Now I suggest that we all go over to the Local Reality Shop, where you can get an alternative and better way of looking at reality."

"But what is randomness?" some heckler in the crowd asks. "You ruled out real nonlocal influences because each record was truly 'random.' "

"Ask the mathematicians," came the reply.

"They don't know what randomness is," says the heckler.

"Neither do I," says the local reality salesman. "But true randomness is unbeatable, which in this case means it will always defeat you if you try to detect real nonlocal influences. There's no randomness like quantum randomness."

The crowd begins to drift out of the Objective Reality Shop toward the Local Reality Shop, led by the salesman, who is feeling rather good after his speech. Someone is asking why he is so critical of the occultists and pseudoscientists, and he begins a little story.

"When I was ten I became fascinated with magic. I learned simple card tricks, built apparatus, and bought magic tricks from mail-order catalogues. The opportunity to perform magic came at friends' birthday parties or holiday occasions, and it was a polished magic show. As a magician and entertainer I responded to the interests of the audience. What struck me was the difference in the response to magic tricks by children and adults. The adults accepted the tricks as entertainment; they wanted to be fooled. Not the children. Their capacity for the suspension of belief was not developed—they wanted to know how the tricks were done. For them it wasn't entertainment; it was a violation of their trust in physical reality.

"A real magician makes no claim to violate physical laws; he only appears to do so. However, when pseudoscientists make claims to discover dramatic new phenomena, going beyond current physical theory, like telepathy or mental metal bending, then, like children, we must insist on seeing how the trick is done or as adults sit back and enjoy the entertainment."

As we enter the Local Reality Shop we see it is already quite crowded with lots of physicists and others who swear by the Copenhagen interpretation of quantum theory only because their heroes, Bohr and Heisenberg, invented it. The salesman who broke up the discussion in the other shop and led us here clearly is the merchant of this one. As the crowd thickens he begins his sales pitch.

"The basis of physics," he begins, "indeed the whole of science, is predicated on the principle of local causality—that material events occurring in a region of space are due to adjacent material events. How can we have a science if an event on the other side of the universe is instantaneously influencing events here now? Quantum theory obeys the principle of local causality. If we accept this principle, then we have to take a hard look at what is meant by objectivity—the assumption that the microworld has a definite state of existence like the macroworld. Scientists are accustomed to thinking in terms of what we actually know to be true about the world, not what we fantasize, and the microworld is a fantasy if we are not actually observing it. Until measurements are actually performed you cannot even talk about the objective properties of things. Physicists all accept that, and I urge you to accept it also."

"But doesn't this imply reality is observer-determined?" someone in the audience questions. "What kind of reality is that?"

"True enough," says the merchant, "but we only have to worry about the observer-determined reality for quantum-sized objects. Of course, quantum-level events influence the macroscopic world —that was the point of Schrödinger's cat—and therefore it seems that the quantum weirdness leaks out into the world of ordinary objects. But that is pushing the Copenhagen interpretation too far, because there is a qualitative difference between the micro-world and macroworld—the macroworld can store information while the microworld cannot. We concluded our discussion of Schrödinger's cat with the realization that the observer-determined reality is only for atomic-sized objects. The reality of these is a distribution of events. By the act of observing we change one random distribution to another random distribution. You can hardly call that an observer-determined reality. It was like those nonlocal influence advocates who turned out to be saying no more than that one random sequence shifted to another."

One distinguished scientist in our group politely asks the salesman how he can be absolutely certain an observation has been made if observation depends on temporally irreversible processes which are themselves only statistical—highly irreversible but not absolutely so. Before the salesman can answer, someone shouts a question at him.

"But suppose there is only one event and not a sequence?" asks the heckler, who has followed the crowd here. "Suppose there is only one event, not a distribution of events, and that event determines whether the human species lives or dies, not just a cat."

"Single quantum events have no significance in quantum theory. They occur at random," says the salesman.

"What is randomness?" asks the heckler. We have heard this before. Our heads are spinning anyway, and the air in the room is very hot. We go out of the shop just as another argument bursts into a shouting match. It is something about consciousness being implied by the Copenhagen interpretation. We never do hear the end of it but are grateful to be outside where the air is fresh. Time for a walk to think things over and clear our heads.

Not far from the reality marketplace we find a cool park, and there, on a bench smoking a pipe, sits an old man whose presence projects both warmth and confidence. "Have you bought a reality yet?" we ask.

"No, not yet, and I'm doubtful I will," he replies in a thick Danish accent. "I have thought about the problem for a long time and have come to some conclusions in discussions with Einstein."

"Where is Einstein now? What reality did he purchase?" we ask our informant.

"Einstein left the reality marketplace a long time ago, leaving his cash to me. He would have none of it and took to wandering farther down the road, like the wanderer he was in his youth. I have no idea what he found there, if anything. As for myself, I have come to terms with quantum reality.

"There is no quantum world like the ordinary world of familiar objects like tables and chairs, and we should stop looking for it. The entities of the microworld like electrons, protons, and photons certainly exist, but some of their properties—basic properties such as their location in space—exist only on a contingency basis. Previous to the invention of the quantum theory, physicists could think of the world in terms of its objects independent of *how* they knew that world existed. Quantum reality also has things—the quanta-like electrons and photons—but given along with that world is a structure of information which is ultimately reflected in how we speak about quantum reality. Quantum measurement theory is an information theory. The quantum world has disappeared into what we can know about it, and what we can know about it must come from actual experimental arrangements—there is no other way.

"What I am certain of is that quantum reality is not classical reality—there is no way you can fit it into classical reality. Quantum theory does not predict individual events and classical theory would; the two theories are logically distinct. But even in our attempt to characterize what quantum reality is *not*, we appeal to classical concepts such as objectivity and local causality. We have no choice in doing this, because we are macroscopic beings and live in a classical, visualizable world to which those concepts apply.

"We can imagine that quantum reality is like a sealed box out of which we receive messages. We can ask questions about the contents of the box but never actually see what is inside of it. We have found a theory—the quantum theory—of the messages, and it is consistent. But there is no way to visualize the contents of the box. The best attitude one can take is to become a 'fair witness'—just describe what is actually observed without projecting fantasies on it. This is a minimalist approach to reality and the one I advocate.

"Those people in the reality marketplace have forgotten something I told them long ago, or perhaps they never heard it properly—the principle of complementarity. This principle asserts that in describing reality we must invoke complementary concepts that exclude each other—they cannot both be true. But not only do they exclude each other conceptually, they depend on each other for their very definition. For example, male and female can be understood as complementary concepts. If you imagine that there is a choice of sex as you are born, then you may pick either female or male. But if the world had only one sex, then there is no concept of sex—the very concepts of male and female define each other as well as exclude each other. Such complementary concepts are different representations of the same single reality—in this example that is the reality of humanity.

"My favorite illustration of complementarity is the picture of a vase made of two profiles used by the gestalt psychologists. Is it a vase or two profiles? You can see it as either, depending on which image is figure and which is background. But you cannot see it as both simultaneously. It is a perfect example of observer-created reality—you decide the reality you are going to see. And yet the definitions of what is the vase and what is the profile depend on each other—you cannot have one without the other. They are different representations of the same underlying reality—here simply a piece of black and white paper.

"Now you know why I stopped going to the reality marketplace. Those two shops for objective and local reality are actually run by two brothers, and other members of the family run the other shops. If you think carefully about the objectivity and locality of the microworld, they turn out to be complementary concepts in the quantum theory—just like the vase and profiles. That is the beautiful feature brought out by Bell's experiment. If you fantasize that the photons exist in a definite state as the flying nails exist in a definite state, then you see that reality must be nonlocal. But the moment you actually try to verify the actual state of a flying photon—which is the same as trying to verify real acausal nonlocal influences—you must upset the first condition of the experiment, which is that the two photon polarizations are correlated precisely. Conversely, if you accept strict local causality then there is no option but to give up the idea of objectivity for individual photons. That is how the principle of complementarity applies to Bell's experiment.

"From the macroscopic view all we have are the records at A and B, and these are certainly objective in the usual sense. Like the live or dead cat they cannot be erased. But the information on these records can never be used to infer real nonlocal or acausal influences. I know there are people who claim the quantum theory requires we give up objectivity or locality for the macroworld of tables and chairs. But they haven't understood that the macroworld and microworld are qualitatively distinct. There is no macroscopic quantum weirdness.

"Arguing whether the microworld is local or objective is like arguing over whether the picture represents a vase or profiles. They are two mutually exclusive ways of speaking about the same reality. You must pick one if you are going to describe quantum reality. But within the framework of material possibilities your reality is a matter of choice. Once your mind accepts this, the world will never be the same again. The material world actually imposed this way of thinking on us. I cannot stop wondering about that. The real mystery of the physical world is why there is no mystery—nothing seems to be ultimately hidden. That we may not always know reality is not because it is so far from us but because we are so close to it."

We feel excited by his remarks, though the old uneasiness has not left us. Yet listening to him is certainly better than that marketplace. After a long silence our old friend gives us his final words. "What quantum reality is, *is* the reality marketplace. The house of a God that plays dice has many rooms. We can live in only one room at a time, but it is the whole house that is reality."

He gets up and leaves us. Only the smoke from his pipe remains, and then, like the smile of the Cheshire cat, that too disappears.

PART II

The Voyage into Matter

God used beautiful mathematics
in creating the world.
—Paul Dirac

1. The Matter Microscopes

Truth, indeed, may not exist; . . . but what men took for truth stares one everywhere in the eye and begs for sympathy. The architects of the twelfth and thirteenth centuries took the Church and the universe for truths, and tried to express them in a structure which should be final.

—HENRY ADAMS, *Mont-Saint-Michel and Chartres*

SOME TIME AGO a friend and colleague, Sidney Coleman, and I were enjoying dinner in a small French restaurant nestled deep in the Jura Mountains near Switzerland. We were visiting CERN, a large international nuclear research laboratory just across the border near Geneva, and like many visiting scientists, we indulged in the cuisine of the fine local restaurants. As the summer sun sank, Sidney sliced into his quenelle, sipped his wine, and we began to speculate on the future of high-energy physics.

Enormous laboratories, like CERN, designed to study the most fundamental structure of matter have been built in the United States, Europe, and the Soviet Union. The main component of these laboratories is a large hollow ring through which protons —quantum particles—are accelerated to very high speed and then collide with a variety of nuclear targets. By examining the results of the collision, physicists learn about the structure of matter.

Sidney and I were theoretical physicists, whose ambition it is to help find a mathematical description of the fundamental structure of matter. But at CERN, the theoretical physicists—although they numbered about one hundred—are but a tiny portion of the total personnel. An even larger number of experimental physicists, drawn to this laboratory from all the universities of Europe and America, compete to use the facilities. Machine builders design and develop the accelerators, while thousands of technicians are employed to build the equipment. Each such lab costs hundreds of millions of dollars and absorbs a good fraction of national budgets for pure research. Sidney and I wondered: Where is the public constituency for high-energy physics research—who cares? Might not research funds be better allocated to areas with more immediate practical goals? I do not believe the answer to such questions can be found in economic analysis. I believe the answer is reflected by the confidence a society has in its idea of civilization.

As theoretical physicists requiring no more equipment than pen and paper for our work, Sidney and I were impressed by the sheer scale of modern high-energy labs—the huge experimental halls filled with hardware, computing equipment, electronic support systems, and measuring devices. There was no question that our field of research had become "big science." But this was not always the case—there was a time when such labs were small. What troubled us was that we knew the last and most decadent stage of any development is giantism. When one cannot think of how to do things better one simply makes things bigger. The construction of the great pyramids in Egypt marked the end of the Old Kingdom. Bigger and bigger cathedrals and temples were built when the faithful became secure and comfortable. Dinosaurs, too, were an evolutionary dead end: the huge reptiles were replaced by small, energy-efficient mammals. Yet some physicists suggest building still larger accelerators in outer space where gravity and space would not limit their size. Are high-energy physics labs, like the dinosaurs, fated for extinction? Is there a better way to study the ultimate structure of matter?

The current answer to this question is no. Once the decision to explore matter down to the finest detail has been made it is clear that the huge size of high-energy accelerators is not gratuitous.

Niels Bohr on a motorcycle. (American Institute of Physics, Uhlenbeck Collection)

Werner Heisenberg and Niels Bohr at the 1934 Copenhagen Conference held at the Niels Bohr Institute. In the foreground is a bottle of Carlsberg beer—the beer that made quantum mechanics famous. (American Institute of Physics, Weisskopf Collection)

Wolfgang Pauli in Carlin, Nevada—a whistle stop on the way to Caltech in California. He was on his way to announce his hypothesis of the neutrino in the summer of 1931. (American Institute of Physics, Goudsmit Collection)

J. Robert Oppenheimer, I. I. Rabi, Harold M. Mott-Smith, and Wolfgang Pauli. (American Institute of Physics).

Richard Feynman and C. N. Yang. (American Institute of Physics, Marshak Collection)

Murray Gell-Mann cutting mushrooms, Tabor Gulch, Colorado, 1974. (Bernice Durand)

A random dot stereogram from Bela Julesz's *Foundations of Cyclopean Perception,* University of Chicago Press. Each square consists of random dots, but the cross-correlation of the two random patterns is not random. To see the three-dimensional image (a square within the square), one must fuse the two parts of the stereogram—not an easy trick, and this should not be attempted by people with poor eyesight. Hold the stereogram about twelve inches from your eyes in good light. Relax your eyes so you begin to go cross-eyed. I have placed two triangles above the squares. First try to fuse them in your visual field. Then look at the fused squares, and after some time you should achieve stereopsis. Once you see the three-dimensional image, the effect is quite dramatic and unambiguous. People who have difficulty with this can examine the random anaglyphs in Julesz's book, which don't require that you go cross-eyed but do call for special viewing glasses.

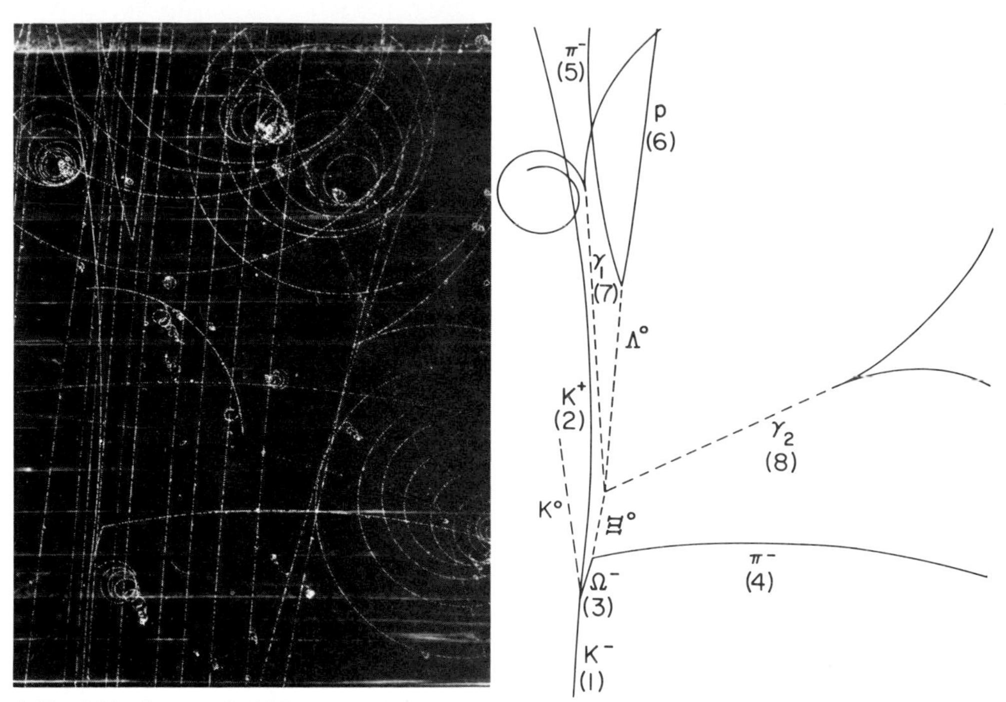

A liquid hydrogen bubble chamber photograph taken at Brookhaven National Laboratory showing the first Ω^- event—a remarkable confirmation of the eightfold way. To the right of the actual bubble chamber photograph is the interpretation of the event with the tracks of the various particles identified (electrically neutral particles leave no tracks but their paths—shown dotted—can be reconstructed). The Ω^- track is the short one near the bottom produced by a K^- meson interacting with a stationary proton (not visible) in the liquid hydrogen.

Albert Einstein sailing, 1936. (American Institute of Physics)

The two-mile-long linear electron accelerator near Stanford University. Part of the interstate highway system runs over the accelerator. Here the quarks inside of the proton were first seen "like raisins in a pudding," and charmed hadrons and their detailed properties were discovered. (Stanford University)

Melancolia **by Albrecht Dürer is a representation of rational inquiry. The magic square on the wall has the property that numbers in each row, each column, and the diagonals all add up to the same sum.**

An aerial view of the Fermi National Accelerator Laboratory. Chicago lies in the background. Protons are accelerated in the huge ring, extracted, and then sent to various experimental areas shown on the left side. A hadron containing the bottom quark was detected here in 1978. (Fermi National Accelerator Laboratory)

A neutrino interaction detector at CERN near Geneva, Switzerland. This large calorimeter actually measures the properties of the elusive neutrino's interactions and gives physicists the clues for building theoretical models of the quantum particles. (CERN)

At first it seems that if we are going to explore the world of very small objects we would need extremely tiny instruments, but just the opposite is true. Big instruments are required because of a curious property of quantum particles. Recall that according to the quantum theory every quantum particle, like the proton or electron, can also be thought of as a little packet of de Broglie–Schrödinger waves. The wavelength of a particle—the distance from one wave crest to the next—is inversely proportional to the speed of the particle. Consequently the faster a particle is moving the shorter its wavelength. If one makes a beam of such particles in a high-energy accelerator, then the smallest object that one can "see" with the beam has to be larger than the wavelength. For example, an ocean wave is not influenced by a swimmer, who is small compared to its wavelength, but it is influenced by a large ocean ship—the wave can "see" the ship but not the swimmer. The wavelength of the particles in a beam is the critical length in determining the size of the smallest object that can be seen using that beam. Hence to detect smaller and smaller material objects we require shorter and shorter wavelengths. The only way to create those essential short-wavelength particles is to accelerate them up to very high speed. And that is precisely the purpose of high-energy particle accelerators.

A high-energy accelerator is essentially a microscope—a matter microscope—that is designed to see the smallest things we know exist—the quanta of elementary particles. The principle of the microscope and an accelerator is the same. In an ordinary table microscope the beam consists of particles of light—photons—which scatter from the object we want to observe under the microscope. Lenses help focus the light to resolve and intensify the image. But an ordinary microscope is useless to examine objects which are smaller in size than the wavelength of visible light. To do that we take the next step to an electron microscope, which uses electrons as the probing particle instead of photons, because the wavelength of even slowly moving electrons is shorter than that of visible light. The electrons can be focused using magnetic lenses which produce magnetic fields to bend the path of the electrons. With the electron microscope and using special methods, one can see down to the level of molecules. To go beyond this level to probe the nucleus of atoms requires yet another technology to accelerate beams of particles to still-higher speeds and correspondingly shorter wavelengths. How can physicists probe the nucleus?

The response to that challenging question marked the modest beginning of the modern high-energy accelerator. John Cockcroft, a young English student of Rutherford's, working in the Cavendish Laboratory in England, suggested in 1928 that protons could be accelerated in an electrostatic field and the resulting high-speed protons could then be used to bombard the nuclei of various atoms. Theoretical calculations based on the new quantum theory showed that Cockcroft's protons ought to penetrate the large repulsive electric fields which acted like a barrier surrounding the nucleus. In 1932, in collaboration with E. T. S. Walton, Cockcroft succeeded in inducing nuclear transmutations with his beam of protons—a sure sign that the nucleus had indeed been penetrated. Humanity had touched the nuclear core of atoms.

In the early 1930s while Cockcroft was doing his experiments in England, an energetic young American, Ernest O. Lawrence, in collaboration with M. S. Livingston, another physicist, had begun to design a new kind of accelerator called the cyclotron in Berkeley, California. Cyclotrons accelerated a beam of particles into a circle, using magnets to bend the path, and in this way the beam could spend more time being accelerated. Lawrence's first machines accelerated protons outward from their source in a spiral orbit between the faces of a large electromagnet. When the particles reached the largest orbit of the spiral they had maximum energy. At first Lawrence was just interested in making high-speed particles and studying their properties. But when he heard of the work of Cockcroft and Walton in England he realized his cyclotron had a new purpose—it could be used as a matter microscope. Lawrence turned his beam of high-energy particles onto nuclear targets. The work of Cockcroft, Walton, and Lawrence marked the beginning of the modern era of experimental nuclear physics. The energy levels of atomic nuclei began to be mapped out.

Lawrence's cyclotrons were enormously successful, and they became the prototype for all subsequent large proton accelerators. Whenever the question arose of building a cyclotron of still-higher energy, the machine builder Lawrence responded, "I can do it." Physicists realized that with the cyclotron they now had an instrument, like Galileo's telescope, that could explore a new realm of nature—the subatomic microcosmos. Beyond the atom, beyond the nucleus, lay a virgin territory, a place never before seen. Physicists believed that in the nucleus lay the clue to the ultimate structure of matter and fundamental laws of nature. But they realized that to go beyond the nucleus and its energy levels and see the very structures that produced the nuclear force, higher energies were required. Again the wavelength of the probing particles would have to become shorter in order to probe the yet smaller subnuclear world.

After the Second World War, physicists returned to the problem of building still-bigger particle accelerators. Many physicists, participants in the Manhattan Project that produced the atomic bomb, had gained experience in the management of large projects under government sponsorship. University professors became administrators, and the symbiosis of government and science research reached a new scale. J. Robert Oppenheimer especially urged government officials that the United States must support scientific research on a broad front, not just weapons development—that the promotion of fundamental research and the practical needs of national security were congenial. Postwar national confidence and economic affluence provided the moral and material support for the building of new and bigger accelerator laboratories in the United States.

In 1945 and again in 1952 there were important technological breakthroughs in the design of the accelerators which resulted in substantial reduction in the size and cost of the magnets. Further, it was also clear that there was no theoretical limit on the energy to which particles could be accelerated—all that was required was more powerful magnets. The first generation of accelerators to implement these new design principles were called synchro-cyclotrons and synchrotrons. The machine builders who discovered these techniques were, like Lawrence, visionaries who wanted to build the instruments that would unveil the ultimate structure of matter, and they now took their place alongside the experimental and theoretical physicists. Victor Weisskopf, a theoretical physicist and statesman of science from MIT, once described the division of labor among physicists this way:

> There are three kinds of physicists, as we know, namely the machine builders, the experimental physicists, and the theoretical physicists. If we compare those three classes, we find that the machine builders are the most important ones, because if they were not there, we could not get to this small-scale region. If we compare this with the discovery of America, then, I would say, the machine builders correspond to the captains and ship builders who really developed the techniques at that time. The experimentalists were those fellows on the ships that sailed to the other side of the world and then jumped upon the new islands and just wrote down what they saw. The theoretical physicists are those fellows who stayed back in Madrid and told Columbus that he was going to land in India.

When the next generation of accelerators—the Cosmotron at Brookhaven National Laboratory on Long Island, New York, and the Bevatron at Berkeley, California—began operating in 1952 and 1954 respectively, the modern voyage into matter began. What was revealed by these new matter microscopes was quite beyond the expectations of the physicists. The subnuclear world opened as a vast uncharted ocean before the eager experimentalists, who, with their beams of high-energy protons, discovered forms of matter never before seen. This new matter, in the form of particles called hadrons, was indeed responsible for the nuclear force. The proliferation of the hadrons was an unanticipated discovery—no one had thought there could be so many new particles, and no one knew what it all meant.

Before the discovery of the hadrons, physicists knew of only a few particles. The electron and the proton were known to exist from the beginning of the century, and in 1932 the neutron was discovered as the other major constituent of the nucleus. Later the pion was discovered, and this particle served as a kind of nuclear glue to hold together the protons and neutrons in the nucleus. These particles and a few others were the only ones known before the deluge of hadrons began to pour out of the new accelerators in the 1950s and 1960s. The new hadrons were given names like the kaon, rho meson, and lambda hyperon, and labeled by letters of the Latin and Greek alphabets. But soon more hadrons were discovered, and physicists, running out of letters, resorted to numerical subscripts and superscripts on their letter names to distinguish the hadrons. There were so many hadrons that theoretical physicists speculated there were actually an infinite number of them.

With the proliferation of the hadrons came a major problem —how to precisely determine the interactions of these new particles. An answer to this problem came in 1953 with the invention of the bubble chamber, a device which greatly facilitated the identification of elementary particles like the hadrons. Liquid hydrogen was put into a chamber and then superheated by very rapidly expanding the chamber. If the liquid hydrogen in this state—right above its boiling point—is disturbed by the passage of an elementary particle, the energy of the charged particle boils the liquid, marking its passage by a line of bubbles in its wake like a stream of bubbles from the bottom of a beer glass. This stream of bubbles is photographed, and from the visible path physicists can determine the properties of the particles. While the technical problems of this device can be severe, it led to direct particle identification.

Of all the fundamental particles, only the proton, electron, photon, and neutrino are observed to be stable. All other particles eventually disintegrate into the stable ones. Some of the new hadrons are metastable, which means that they live just long enough so that one can see their tracks in the hydrogen bubble chamber.

However, most hadrons are extremely unstable and will leave no tracks. But from the hadrons that do leave tracks in the bubble chamber physicists can deduce the existence of the extremely unstable ones. Over the next decade, bubble chambers were responsible for the discovery of many new hadrons called mesons, hyperons, strange particles, excited states of the neutron and proton, hyperon resonances—a good part of the whole hadron zoo.

The bubble chamber exemplified the new technology that the construction of high-energy particle accelerators promoted. The new realm of the microcosmos challenged scientists to develop high-speed electronic systems for counting the particles, new high-vacuum technology, superpowerful magnets, and many other ingenious devices.

But what were the hadrons—the thousands of new particles seen in the experiments—telling us about the ultimate structure of matter? It was clear that these new particles represented a new level of matter. The proton and neutron, the primary nuclear constituents, were now understood to be but two particles, distinguished by their relative stability, out of thousands of other less stable hadrons. Physicists had hoped that building the matter microscopes and exploring the smallest things would show matter to be simpler, not more complicated. The proliferation of the hadrons seemed to deny that hope. Was nature playing a cruel trick on those physicists who held to the view that nature would reveal its simplicity as they explored the smallest distances?

By beginning on the voyage into matter, physicists discovered the voyage does not end with atoms or their nuclei but opens into the vast sea of hadrons. They found a new frontier in the world of subnuclear matter which, like the frontier of outer space, seems endless. It is as if we are traveling across a great ocean, the end of which is lost in darkness. Can we eventually land our ship, or like the *Flying Dutchman* are we doomed to sail forever? The problem of interpreting the vast number of hadrons and making a meaningful picture of nature preoccupied the theoretical physicists in the 1960s. They asked what the ultimate structure of matter could be; what the logical options are that were consistent with the laws of quantum physics and experimental observations. Physicists found only three logical possibilities for the end of the voyage into matter.

The first possibility we might call "worlds within worlds." Each time we reach a certain level of matter at very small distances we discover that the world of matter on that level is not elementary and indivisible. Atoms were once thought to be indivisible. But physicists split the atom into its parts. Then the parts were discovered to have yet smaller parts, and so on into an endless regress. The levels of matter could be inexhaustible—it is possible that there are no truly "elementary" particles, that all particles are composites of yet smaller particles.

Alternatively there may be a "rock bottom" to the levels of matter, truly elementary particles which cannot be subdivided. All other particles would be composites of these elementary particles, the ultimate physical stuff—a stop sign on the voyage to ever shorter distances.

The third possibility goes by the name of the "bootstrap hypothesis," inspired by the Baron von Münchhausen, a teller of tall tales who reputedly lifted himself up by his own bootstraps. According to this hypothesis, there may be a level of matter which is both elementary *and* composite, and matter might "lift itself up into existence by pulling on its bootstraps." The basic idea is that we may reach a level of matter for which the fundamental particles are indeed composite. But what they are made out of is more particles of exactly the same kind. This puts an end to the "endless" regress, because although particles at this level are divisible they are not divisible into new kinds of particles. It would be like a pie which if cut in half, would result in two pies identical to the first. No way of cutting the pie would result in smaller pieces, just more pies.

Of these three possible answers, the worlds-within-worlds and rock-bottom possibilities were already thought of by physicists before this century. The bootstrap hypothesis made sense only after the new physical laws of quantum theory and relativity theory were discovered in this century. According to these theories,

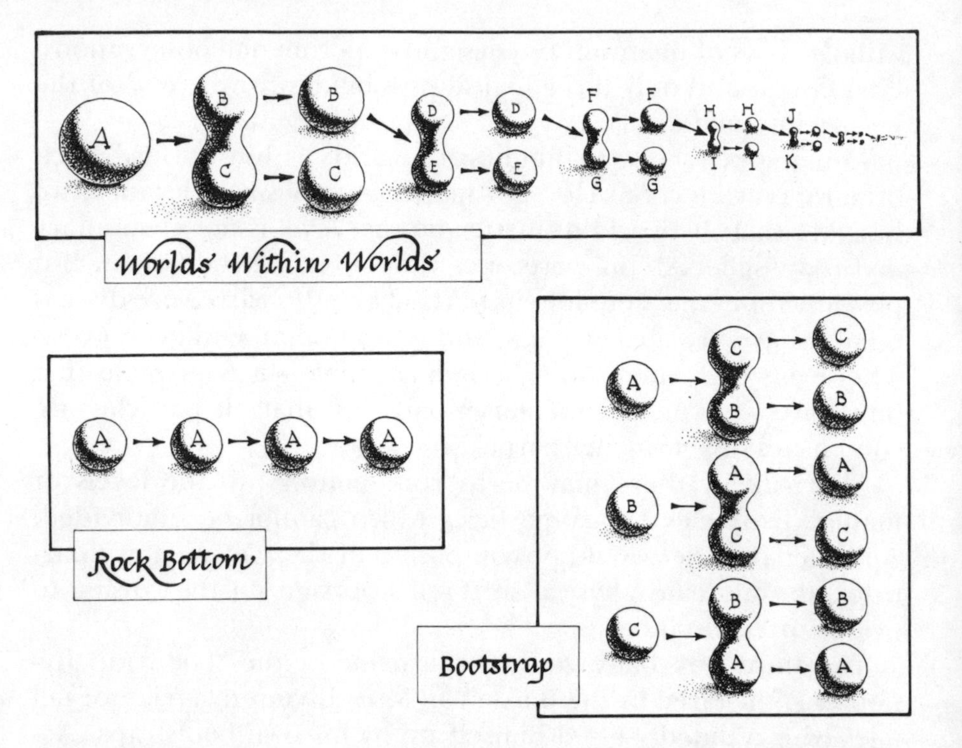

Three scenarios for the voyage into matter. Worlds within worlds: Particles are infinitely divisible into different smaller particles. Rock bottom: A level of matter is reached for which further subdivision is impossible—a true elementary particle. Bootstrap: Particles can be subdivided but the results are the particles that one started with.

the concept of a particle underwent a change which made the bootstrap hypothesis a physical possibility—particles could be both composite and elementary, a possibility previously unthinkable. The lesson is that the search for the ultimate structure of matter is conditioned by the laws of physics. Conceivably, as we probe to smaller distances the laws of physics will be different and new scenarios for the structure of matter will become thinkable.

During the early 1960s when new hadrons were still being discovered, many physicists were attracted to the bootstrap hypothesis because it implied that all these perplexing hadrons were actually made out of one another, so there wasn't anything beyond hadrons. If one smashed hadrons together, as was done in the high-energy accelerators, all that emerged in the pieces was more hadrons, never anything that could be identified as a more elementary particle. Cut apart a hadron and you find more hadrons. Maybe the vast number of hadrons was actually the sign that physicists had arrived at the end of the voyage into matter.

As attractive as the bootstrap hypothesis is, it is not widely held today. Most physicists maintain that hadrons are composite objects made out of more fundamental particles called quarks. How did this change in outlook come about?

In the early 1960s, theoretical physicists, led by Murray Gell-Mann, discovered that the hadrons organized themselves into families or classes. This organization principle was called the eightfold way.* It was based on a mathematical symmetry and was extremely successful in correlating the experimentally observed properties of the hadrons. An easy way to understand why the eightfold way worked was to assume hadrons were actually made out of new, smaller particles, the quarks, only three of which were needed to build all the hadrons. Each hadron could be viewed as made up of a few quarks orbiting around each other in a specific configuration. Since the quarks could orbit in an infinite variety of different configurations, there was actually an infinite number of hadrons. This quark model was invented by theoretical physicists in order to simplify the complexities of the hadrons—and it worked. But no one had ever seen a quark. The central problem now became: Where were the quarks and how could one experimentally confirm the quark model?

Even before theoretical physicists imagined the existence of quarks, experimental physicists at Stanford University had turned up evidence that the proton—the first-known hadron—could not be a pointlike particle without extension in space but

* The original "eightfold way" comes from an aphorism attributed to Buddha: "Now this, O monks, is noble truth that leads to the cessation of pain: this is the noble Eightfold Way: namely, right views, right intention, right speech, right action, right living, right effort, right mindfulness, right concentration."

instead actually occupied a volume in space; it was like a tiny ball, not a mathematical point. They shot a beam of high-energy electrons at a proton target and by examining the results of the collisions, concluded that the proton (and by implication all hadrons) had a definite structure and extension in space. With retrospective wisdom we now know this discovery provided the first hint that the hadrons were not elementary.

Encouraged by these results, a much larger version of this electron accelerator was built near Stanford University in the mid-1960s. Here physicists conducted the crucial experiments which confirmed the idea that hadrons were made out of new tiny particles—the quarks. The new machine, over two miles long, is located at the Stanford Linear Accelerator Center (SLAC). Its basic design is completely different from the synchrotons because it is not even constructed in a ring. Instead, its basic structure is a two-mile-long linear vacuum pipe down which electrons are accelerated. The energy is supplied to the electrons by a series of klystrons, devices which pump microwave energy in the form of electromagnetic waves into the pipe at regular intervals. The bunched electrons ride the resulting electromagnetic wave down the two-mile pipe like a surfer on an ocean wave. At the end of their ride the electrons have an enormous energy and then collide with a variety of targets in the experimental area.

Critics of this machine argued that electrons—which had primarily electromagnetic interactions—could never teach us about hadron structures and insisted one needed proton beams to study such strong interactions. The critics were wrong. When the structure of hadrons was determined, the quarks inside the hadrons could be quite precisely probed by the electromagnetic interaction of the electron, not the strong interaction of the proton. Ironically, if quarks had not existed inside of hadrons, the hand of the critics would have been strengthened. The SLAC experiments confirmed the standard picture of the quark model: Hadrons were built out of quarks bound together by strong forces.

In the late 1960s, as a result of the electron scattering experiments and other experiments carried out with neutrino beams, it was clear that a new level of matter had been discovered—the quarks. All the hadrons, the great zoo of subnuclear particles, could, it seemed, be built up out of just three quarks. This was an immense simplification emerging out of the chaos of the hadrons. But how does one study the quarks or discover more of them if they exist? For this a new accelerator technology had to be developed—the electron-positron colliding beam machine, a new matter microscope that probed even finer distances. These machines actually collided matter in the form of electrons with antimatter in the form of antielectrons called positrons. Bringing matter and antimatter into contact results in a spectacular annihilation into lots of different hadronic particles. Consequently this new method of counterrotating beams of matter and antimatter and letting them collide was very effective in creating new forms of matter.

The first results of this new technology came from a colliding beam machine at Frascati, Italy. Then a similar machine but with much higher energy was built at the Stanford Linear Accelerator Center. In November 1974, experimentalists at this colliding beam machine and also at Brookhaven National Lab announced to an astonished community of physicists the discovery of a remarkable hadron made out of a new fourth quark. Subsequently more hadrons built of the fourth quark were discovered at Stanford and confirmed at a German laboratory near Hamburg—perhaps the most convincing evidence for the correctness of the quark model for hadrons. Then in 1978 yet another hadron, undoubtedly made out of a new more massive fifth quark, was discovered at the Fermi National Accelerator Laboratory near Chicago. The detailed properties of these hadrons built out of the fourth and fifth quarks were studied by physicists using the colliding beam machines. These now became the new ships upon which to voyage into the realm of the quarks.

Many theorists believe at least a sixth quark must exist at still higher energy, but that discovery lies in the future. New high-energy accelerators are being built. At Brookhaven, ground has been broken for a storage ring for protons. At CERN near Geneva, an immense proton-antiproton colliding ring may be working in the early 1980s as well as an immense electron-positron colliding ring. A proton synchrotron is planned at Surpukov in the Soviet Union, while Japan is also constructing a new accelerator. The United States, which once completely dominated high-energy physics, has become an equal partner with the other industrial nations in the exploration of matter.

New physics of the most fundamental kind will come from these machines. Perhaps new quarks or other exotic particles will be discovered. The proliferation of the quarks has startled physicists. Not long ago only three quarks were required to build the hadrons, and now there are five, perhaps six. Some physicists think too many quarks exist for the quarks to be truly elementary. Although the relatively small number of quarks are a great simplification over the infinite numbers of hadrons, most physicists are not yet satisfied.

Are quarks made out of yet more fundamental structures? Although there are suggestions in this direction, no real progress has been made. The quarks appear to be point particles without further structure—a "rock bottom" of matter. Some theoretical physicists think that beyond our present energies there is only empty ocean. But no one has a clue as to the correct theory that might explain the observed masses of the quarks—it is a complete puzzle. The fact that such puzzles about the quarks remain unanswered indicates we have yet something to discover. If puzzles of the past provide any guide to present puzzles, then the clue to their solution will have to come from experimental exploration at yet higher energy. But what have we learned so far?

If we look back over the century we can see how far we have come in our voyage into the realm of matter. Five distinct levels of matter have been identified: the molecules, atoms, nuclei, hadrons, and quarks. In order to understand any level it was necessary to penetrate deeper. If anything, the structure of matter at the deeper levels seems to get simpler. Eight dozen atoms were a simplification over their millions of molecular compounds. The nuclei of the eight dozen atoms were bound states of just two hadrons, the proton and neutron. With the proliferation of the hadrons in the 1960s it appeared that matter was becoming more complicated. But the quarks—the next level—brought order into the realm of the hadrons. Here we stand today.

In the future, additional levels may appear—perhaps governed by new laws of physics. We have no evidence that anything new must happen; it seems for all purposes that with the discovery of quarks we have reached the end of our journey. But there is an uneasy feeling among physicists that the trip is not over.

Physicists found a cosmos inside of matter. Using high-energy accelerators—the matter microscopes—they made contact with the invisible realm beyond the atom and determined its laws. What we learned there is teaching us about the origins of the universe, that time billions of years ago when everything consisted of an expanding fireball of the ultimate constituents of matter. Eventually we shall become masters of the realm beyond the nucleus and practical devices utilizing the new physics will serve human ends. What these devices will be we cannot yet know, but if the past relation between fundamental research and technology is any indicator, we are in for some remarkable developments.

All the detailed knowledge about the subnuclear world began in great accelerator laboratories, the ships upon which we have voyaged into the universe. These laboratories can be compared with the cathedrals built in the age of faith in Europe. The construction of the cathedrals served as a focus for the energies of the best architects and craftsmen, challenging their genius and skills to produce new techniques. These great spaceships of faith, designed to convey the souls of men to God, inspired their builders to span great spaces, achieve new heights, and invent new materials. But providing new technology was not the purpose of building the cathedrals.

To advance civilization, a vision is needed. The cathedrals of Europe were built by a people consumed by a vision of faith. But reason also has its dreams. The great scientific laboratories implement our contemporary dream to solve the puzzle of the cosmos. Perhaps in some future time when people look back on our age they may no longer share our sense of truth, and yet just as we are moved by the vision that built the cathedrals they will be touched by our vision—the vision that knowledge is the ultimate instrument of human survival in the universe.

2. Beginning the Voyage: Molecules, Atoms, and Nuclei

On the occasion of one of Rutherford's discoveries, Arthur Eve, a colleague of his, said to him: "You are a lucky man, Rutherford, always on the crest of the wave!" To which he laughingly replied, "Well! I made the wave, didn't I?"

SOMETIMES I GET the feeling that I am living in a three-dimensional movie. The movie began billions of years ago with the big bang that created the universe and has been going on ever since. Everything in the universe, the stars, the sun, the earth, my body and yours, are part of the scene. We are all in the movie; it is the only show in town. It is not clear where the dramatic action is leading or if the movie has a plot or a director. As a physicist my interest is in how the scenery for this 3-D movie is constructed, what its parts are, and how the machinery works. I certainly didn't ask for it, or even to be in it; nevertheless here we all are in the cosmic movie.

Existential philosophers speak of being "thrown into the world," but I prefer the psychiatric descriptions of "dissociation." Human beings can dissociate their minds from the world and create metaphors like the 3-D cosmic movie. Metaphors are freely created symbols transcendent to the world which, if we share them, organize our social experience in new ways. That is the great and sometimes dangerous power of symbols. With these warnings aside, let us indulge the idea of the universe as a 3-D movie and ask, "What is it? Who ordered that?" Physicists are asking just those questions.

To answer such questions, theoretical physicists turn to the clues provided by the matter microscopes, and with some inspired guesses they attempt to devise their version of the great 3-D movie. In this century physicists have peeled through five levels of matter like the layers of an onion—the molecular, atomic, nuclear, hadronic, and quark level. Each was discovered through new experimental techniques using matter microscopes probing ever shorter distances. By precise measurements of these five levels of matter, physicists learned the properties of the interactions that have produced them.

Our purpose in this and the next several chapters is to explore the five levels of matter—molecules, atoms, nuclei, hadrons, right down to the quarks. These objects are the cast of characters, the actors, in the cosmic 3-D movie. Following the presentation of the cast comes the movie itself—the interactions among our actors, which are mediated by yet another set of quanta called gluons. Finally, we turn to the analysis of the play of particles and what it might mean for forming a picture of material reality. Someday, experimental physicists may find new kinds of matter beyond the quanta at smaller distances yet. Then we may have to rethink our analysis. But within well-defined experimental limits nothing has been overlooked. That is the message of the physics of the last few decades and the challenge confronting us. If nothing is hiding, physicists have got to account for what there is. What excites physicists is that in the last decade, after many detours, an order has emerged. We seem to be on the verge of a new synthesis, a unification of all the forces of nature. That grand synthesis and its implications for the origin of the universe will be the topic of the final chapters of our voyage into matter.

MOLECULES

Our voyage into matter begins with molecules—the basic substances out of which everything we see, hear, smell, touch, or taste is made. The smallest quantity of anything that you can have is a single molecule, because if you divide a molecule into its constituent atoms it ceases to have its chemical properties—it can become a highly unstable substance with different properties. Water is a molecule made out of two hydrogen atoms and an oxygen atom. At room temperature, water is a liquid but hydrogen and oxygen are gases—split a molecule like water into its atoms and it just isn't the same.

Molecules range in size from a few atoms like the water molecule to complex arrangements of tens of thousands of atoms in large organic molecules. We can think of the atoms as the basic construction materials for the molecules, which are the buildings and machines of the microworld. The rules for putting the atoms together into molecules are known from the quantum theory, but the interactions of the molecules with themselves is a complex study in its own right. Some molecular structures are highly repetitive arrangements of atoms resulting in crystals or metals—a rigid, fixed framework. Other molecules have freed their bonds to their neighbors and freely move around in the random motions that constitute a gas. Molecules that have only partially freed their bonds may slip and slide over each other and make up liquids. There are literally billions of different kinds of molecules that can exist at the temperatures typical of our planet's surface, and this great variety is reflected by the vast variety of substances we find in our world. At higher temperatures or lower temperatures than that on earth the number of different molecules and their interaction possibilities are greatly reduced, with a corresponding decline in complexity. We are at just the right temperature range for maximizing molecular complexity, a complexity on which life itself depends.

Chemists and molecular biologists are the architects and engineers of the molecular microworld who work out the methods for combining molecules. In this work the advent of computers, which manage large chunks of data better than humans, have been helpful in determining the structure of the larger molecules. Exploration of the world of very large molecules is only beginning, and I'm sure there are surprises in store for us.

In 1959, Richard Feynman, a physicist at the California Institute of Technology, gave a talk titled "There's Plenty of Room at the Bottom." His remarks may well prove to be prophetic. Feynman sees the world of molecules as a potential building site for all sorts of new structures where we could build tiny devices that would perform specific tasks. Molecular-sized repairmen could be released in the human body and go to a damaged site to fix it. We can imagine building tiny "cities" with industries that produce specific molecular-sized instruments. Infinitesimal computers could control this molecular world—the frontier of miniaturization. Molecular "societies" could be constructed for human ends. The microworld is a realm as vast as outer space, and human mastery of that world is just beginning. Conceivably, the survival of our civilization could depend on our ability to master that microworld.

The molecular architects are only beginning to imitate nature in their molecule building—and imitation is the sincerest form of flattery. If we turn to nature as a guide to what can be done we see that the most beautiful and certainly the most complex molecules are the organic molecules used in the processes of life. This natural molecular architecture is the work of hundreds of millions of years of molecular evolution.

Most people find evolution implausible. Why is my spine erect, my thumb opposable? Can evolutionists *really* explain that? Once I attended a lecture given by the writer Isaac Bashevis Singer, and one of many biologists in the audience asked Singer about evolution—did he believe in it? Singer responded with a story. He said there was an island upon which scientists were certain no human being had ever been. When people landed on the island they found a watch between two rocks—a complete mystery. The scientists when confronted with the evidence of the watch stuck to the view that the island was uninhabited. Instead they explained that although improbable, a little bit of glass, metal, and leather had over thousands of years worked its way into the form of a watch. Singer's view differed from that of the scientists—as he summarized, "No watch without a Watchmaker." This story reflects the feeling many people share that random chemical interactions cannot explain the existence of life on earth. The reason it is hard for such people to grasp the evolutionary viewpoint—and for this our feelings do not help—is the difficulty in grasping the immense time a billion years actually is.

According to the view held by scientists, simple molecules, in an appropriate environment, will combine to form more complex molecules that can automatically replicate themselves. How that happened in the ancient oceans is still the subject of scientific speculation—but there is no reason why it can't happen. Those ancient molecules, of which RNA and DNA—the genetic material—are the modern progeny, were the first material basis for life. Once upon a time there was a single molecule on this planet with the self-replicating property, a property no other molecule previously had. It must have made billions and billions of copies of itself in an orgy of reproduction. It probably didn't stop replicating until an error occurred that produced a different self-replicating molecule—its own competition and the beginning of molecular evolution. In the words of the physicist Gerald Feinberg, life appears to be simply "a disease of matter."

If we did not have such good evidence for it, evolution would be a completely implausible plot in the 3-D movie. Who could ever have imagined that from the war of nature, from famine and death, the highest and most exalted forms of life evolved? The idea of evolution is too implausible to have been *imagined.* It had to be *discovered* through careful observation of the natural world.

One could devote a worthwhile scientific career to the grand study of the architecture of molecules. Aided by computers and other new technology, scientists will fashion new molecular buildings, remarkable for their function and purpose, in the next century. The microworld of molecules is a new frontier which has only begun to be exploited. As exciting as this frontier is, as explorers on the voyage into matter we must leave it and journey onward deeper into matter. Our next step takes us to the building material of the molecules—the approximately eight dozen different atoms out of which they are all made. And with the atom we enter the world of quantum weirdness.

ATOMS

Ernest Rutherford in 1911 made the first experimental determination of atomic structure—the first major step in understanding the atom. Rutherford found that individual atoms have a tiny positively charged core—the nucleus—only one ten-thousandth the size of the whole atom. The size of the atom was determined by the relatively larger cloud of electrons that swarm around the tiny nucleus. Almost all of the mass of the atom—and hence of ordinary matter—is concentrated in the tiny nucleus; the swarm of electrons weighs almost nothing. However, it is the properties of electrons in their orbits about the nucleus that determine the interactions between atoms and hence the laws of chemical combinations for the formation of molecules. Discovering those chemical laws from an atomic theory took over two decades after Rutherford's work.

Two great steps in atomic theory were required before these chemical laws could be found. The first step was taken by Niels Bohr, who applied the old quantum ideas of Planck and Einstein to Rutherford's model of the atom. Bohr's theoretical model of the atom successfully explained the light spectrum of the hydrogen atom but raised many questions of principle about the application of classical physics to atomic systems. The second great step was the invention of the new quantum theory, which overthrew classical physics, provided the mathematical basis for a complete understanding of the properties of atoms, and initiated a revolution in understanding material reality.

When we view this scientific achievement from the perspective of half a century we can see there was a "gift of nature" that facilitated progress—the existence of the hydrogen atom. Hydrogen is the simplest atom, consisting of a single proton as the nucleus with a single electron in orbit about it. Because nature gave physicists such a simple system on which they could test and work out their ideas, rapid progress in discovering the laws of the atom could be made. The spectrum of light emitted by hydrogen is regular, and this regularity was elucidated by Bohr's model. Imagine how much more cumbersome it would have been if the simplest atom was oxygen, with its swarm of sixteen electrons and a correspondingly complicated light spectrum. Then it might have taken physicists hundreds of years to discover the laws of the quantum theory which described the motion of all those electrons. Instead, by deducing the general quantum laws that worked for the simple hydrogen atom they developed the confidence that these same laws worked for more complicated atoms.

Such a "gift of nature" was bestowed once before over two centuries ago, and resulted in the discovery of the law of gravitation. The gift of nature to the physicists—or natural philosophers, as they were called then—was the simplicity of the solar system. Suppose that the earth did not have a moon, as Venus or Mercury do not, and instead of only one sun we had two or three. The orbit of the earth would then be quite complicated. Nothing like Kepler's first law, which stated that the planetary orbit was an ellipse with the sun at one of the foci, would have been easily found, and we might still be searching for the law of gravity. These gifts of nature show how progress in understanding the universe is profoundly conditioned by the existence of a simple system in the environment. Conceivably, the lack of such simple systems in other areas of science has inhibited progress.

With the invention of the quantum theory, progress in understanding the properties of atoms was rapid. The laws of chemistry could now be put on a more fundamental basis—they were the consequence of the quantum interactions of the electronic swarm surrounding the nucleus. The weird reality implied by the quantum theory was right. Aided by these new ideas, atoms became the playthings of the experimental physicists.

The atom has two basic components, the cloud of electrons and the nucleus at its center. While some physicists became interested in studying the electronic properties of atoms, others turned to examine the nucleus, the heart of the atom.

NUCLEI

At first the nucleus, the other primary component of the atom, was a complete puzzle to physicists—a big question mark at the center of the atom. They knew that the tiny nucleus accounted for most of the mass of atoms, and it was a place of enormous energy transformations, as evidenced by radioactivity—the emission of particles from the nucleus. Then in 1932, Chadwick, the English physicist, discovered another major particle he called the neutron, which was similar to the proton in its properties except that it had no electric charge. With this discovery it was clear that atomic nuclei were composed of two major particles—protons and neutrons—which were bound together into a tiny region by immense nuclear forces. But not until the nuclear matter microscopes, especially the cyclotrons, became available did the nucleus begin to reveal its secrets.

By probing the nucleus with high-energy beams of particles, physicists determined that the protons and neutrons in the nucleus had a definite organization—they arranged themselves into shells. The protons and neutrons could jump from shell to shell much as the electrons jump from orbit to orbit and thus release energy. The nuclei of the eight dozen atoms each had a characteristic set of energy levels. The protons and neutrons shifting their positions inside each nucleus like a restless sleeper signaled transitions from one energy level to another. The observed energy levels of the nuclei were hundreds of times greater than those of the electrons swarming around the nucleus. What this meant was that the force that bound the protons and neutrons together into a tight little nucleus was hundreds of times stronger than the electric force field that was known to bind the electrons to the nucleus. What could produce such a strong nuclear force?

H. Yukawa, a Japanese theoretical physicist, addressed this question, and he suggested an answer based on an analogy. He reasoned that if the electromagnetic force which bound electrons to the nucleus had an associated quantum particle—the photon or particle of light—then the nuclear force should also have an associated quantum particle, one which had to have a very strong interaction with the protons and neutrons, hundreds of times stronger than the photon's interaction with the electrons. Further, the nuclear force associated with this new particle had to act strongly over only a very short distance, because the nucleus was so tiny. The conclusion of his reasoning was that the nuclear force was due to a new massive particle with strong proton-neutron interactions. It was called the meson, a term which was eventually used to denote a whole family of subnuclear particles.

In 1946, Yukawa's theoretically postulated particle was experimentally discovered in cosmic rays—a natural source of high-energy particles. His reasoning was vindicated: Nuclear forces had associated quantum particles. The new particle, denoted by the Greek symbol π, was dubbed the pi-meson or pion for short and had just the mass and interactions required for Yukawa's meson. A few years later, in 1948, when new cyclotrons were built, pions were artificially created in the laboratory—real pions were flying out of the nuclear reactions induced by beams of high-energy particles. However, the most exciting discovery from the new cyclotrons came around 1952—the discovery that the proton and neutron could be energetically transformed into a new state of matter, the nucleon resonance.

What is a nucleon resonance? A way of describing this new state of matter is to make an analogy with a guitar string. Such a string, if unplucked, has a certain energy—the lowest energy of the string. In our analogy with nuclear physics, such a state of lowest energy corresponds to the proton and neutron. If I supply energy to the string by plucking it, the string will resonate and thus have a higher energy. Similarly, if I supply energy to a proton or neutron, they can be energetically excited to become a nucleon resonance. There may be higher energy states corresponding to overtones or harmonics of the guitar string. These higher-energy states have also been observed for protons and neutrons—additional nucleon resonances. The nucleon resonance was extremely unstable and rapidly disintegrated into either a proton or neutron plus a pion; but it was unambiguously there.

The meaning of this discovery was quite profound. What it meant was that the proton, neutron, and pion were not alone. Physicists had discovered new quantum states of matter with an energy level hundreds of times greater than any previously seen in the nucleus. The nucleon resonance and the pion were simply the first to be observed. These new particles were called hadrons —from the Greek word for "strong"—because they all had strong nuclear interactions.

Indeed, when the next generation of accelerators, the Cosmotron at Brookhaven National Laboratory (1952) and the Bevatron at Berkeley (1954), went into operation many more new hadrons were discovered. These machines with their high-energy beams of protons could create new forms of matter never previously seen—something that could not have been anticipated in even the wildest dreams of reason. The voyage into the ocean of hadrons had begun.

3. The Riddle of the Hadrons

Everything that is not forbidden is compulsory.
—MURRAY GELL-MANN

IN THE EARLY 1950s the cast of characters—the fundamental quanta—in the 3-D movie that physicists studied was rather small. This was soon to change. The new matter microscopes—accelerators built in the 1950s and 1960s—used high-energy beams of protons to bombard other protons in a liquid-hydrogen bubble chamber. According to Einstein's relativity theory, energy can be converted into matter—and that is just what the energy of the protons in the beam did when it collided with other, target protons. The energy in the beam created the new forms of matter called hadrons. The proton, neutron, and pion—the first hadrons—were just the tip of the iceberg. Today physicists believe that there are an infinite number of hadrons, most of them highly unstable, disintegrating in less than a billion-billionth of a second into more stable hadrons. Enrico Fermi, the Italian-American physicist, as he witnessed the proliferation of the hadrons, commented that had he known this was to become the outcome of nuclear physics, he would have studied zoology.

Fermi was expressing the disappointment of many physicists that exploring beyond the nucleus did not make the subnuclear world simpler. Smashing a high-energy proton beam on other protons, as was done at the accelerator laboratories at Berkeley and Brookhaven in the United States, CERN near Geneva, and Dubna and Surpukov near Moscow, did not reveal a simpler structure. Instead, all those hadrons were produced. What could be the meaning of the proliferation of hadrons? This was the riddle that confronted high-energy theoretical physicists in the 1960s.

How did theoretical physicists think of hadrons? It became clear from experiments in the early 1960s that hadrons had a definite extension in space—unlike the electron, which seems to behave like a mathematical point particle. Hadrons, like the proton, might be visualized as little balls of bound energy with no observable structure inside them. The hadron-balls could spin and they had electric charge and magnetic properties—but their interior was a blank, a terra incognita.

The fact that the hadrons could spin, like tiny tops, led to the first principle of their classification. The spin of a hadron—like that of all quantum particles—was quantized and could take on only certain values such as 0, 1/2, 1, 3/2, 2 . . . , an integer or half-integer value in certain units. The existence of this quantized spin led to a classification of the hadrons into two great subsets—the set of "mesons," which had integer spin 0, 1, 2 . . . , and the set of "baryons," with half-integer spin 1/2, 3/2 . . . . The proton and neutron with spin 1/2 are examples of baryons. The pion with spin 0 is a meson. Every hadron is either a meson or a baryon.

The distinction between the mesons and baryons is very important, because they behave differently in hadronic interactions. The number of baryons going into a collision is equal to the number of baryons leaving the collision—the law of baryon number conservation. By contrast there is no law of meson number conservation. Hadron collisions can be profligate in creating mesons.

Many theorists focused their attention on the observation that the strong interactions of the hadrons exhibited certain new conservation laws like baryon number conservation and the absolute conservation of electric charge. Hadrons, besides possessing electric charge—the proton, for example, has one unit of electric charge—had other new kinds of discrete charges which were conserved when hadrons interacted. These new charges were given names like "isotopic" charge or "strangeness" charge. Hadrons, when they collided, producing yet more hadrons in complicated interactions, always preserved the exact amount of electric, isotopic, and strangeness charge—the same charge existed after the collision as before. There was no deep understanding of why hadron collisions preserved these charges—it was simply observed experimentally.

Physicists had witnessed something like these laws of charge conservation long ago in a different area of science—chemistry. Chemical reactions among molecules can be extremely complicated, like hadron reactions, but as every student of elementary chemistry knows, the number of atomic elements of a specific kind that enter a reaction must also leave. For example, two atoms of hydrogen plus one atom of oxygen can combine to form the water molecule. At the beginning of this chemical interaction there were two hydrogen atoms, and in the end, bound up in the water molecule, there are still two hydrogen atoms. The same amount of hydrogen, oxygen, carbon, iron, and so on that enters a chemical reaction must leave. This conservation of atoms in molecular reactions is similar to the conservation of the various charges physicists discovered by observing complex hadron collisions. These new charge conservation laws provided an enlightening clue about the structure of hadrons—but that story is for our next chapter.

In summary, the main features of any hadron that physicists knew about were its mass, its spin, which classified it as either a baryon or meson, and the amount of each one of the various charges it carried. These features were crucial in classifying the various hadrons, the first step toward bringing order to the chaotic world of hadronic particles. Physicists could now make tables of the hadrons they had discovered, with each hadron fitting into some entry in the table. Hadrons were classified much in the way that atoms corresponding to the chemical elements were classified in the periodic table.

In 1961, Murray Gell-Mann, a physicist at Caltech, and independently Yuval Neeman, an Israeli intelligence officer turned physicist, noticed a pattern in the already classified hadrons. They based their research on a mathematical symmetry which incorporated the known conservation of the various hadronic charges. But the mathematical symmetry, which implied the pattern they called the eightfold way, went far beyond the conservation laws it incorporated. According to the eightfold way, every hadron must be a member of a specific family of hadrons. These families consisted of a definite number of members; the smallest had 1, 8, 10, and 27 members. Families containing only a single member (hardly a family!) are called singlets; other families with 8 members are called octets, with 10 members, decuplets. All hadron members of a specific family have identical spin, but their electric, isotopic, and strangeness charge differed.

The eightfold way was a big step toward answering the riddle of the hadrons. The proton and neutron were now seen as just two members of a larger family consisting of eight particles called the baryon octet. What were the other six particles? Even before the eightfold way was discovered, these particles had already been detected in the accelerator laboratories. Physicists assigned Greek letters to these six new hadrons—the Λ (lambda) particle, three Σ (sigma) particles, and two Ξ (xi) particles. Likewise, the pion was now also understood to be a member of another family of eight particles called the meson octet. Fitting the hadrons into families was the classification principle of the eightfold way, and it worked beautifully, just like the periodic table of atomic elements. Many properties of a given family of particles, such as their different masses, could now be related using the mathematical symmetry. Exploring these and many other consequences of the eightfold way preoccupied physicists in the mid-1960s.

Some critics of the successes of the eightfold-way symmetry thought its success might be accidental. After all, they argued, it only explained properties of the hadrons already known from experiment, and maybe the theory was just set up to fit the facts and do no more. But there was a new prediction of the eightfold way—the existence of a new particle called the omega minus (Ω^-) postulated by Gell-Mann. This particle, never before seen, would if discovered convince the critics. How did Gell-Mann know the Ω^- had to exist? According to the eightfold way there was a family of hadrons with ten members called the decuplet. Seven of these ten members of the decuplet could be identified with known hadrons. But there was still a gap of three missing members in 1962, when Gell-Mann attended a high-energy physics conference held at CERN. One of the results experimental physicists announced at the conference was the routine discovery of two new hadrons. Gell-Mann immediately saw the connection with the eightfold way. These two hadrons had just the right properties for the decuplet; when added to the seven others that were already known, the resulting set comprised nine particles of the ten. This meant that the tenth particle, the Ω^-, had to exist. Nature should not exhibit a partial regularity without going all the way.

In November of 1963, a large group of experimental physicists at Brookhaven Laboratory devoted its resources to the search for the Ω^-. They took over fifty thousand bubble chamber photographs, and on one of them appeared the Ω^- track. That December the elated experimentalists sent out a season's greeting card which was a bubble-chamber picture of a track left by the Ω^-. When the Ω^- was discovered at the predicted mass value, all but the most hardened critics were won over to the eightfold way.

By the mid-1960s, the eightfold way and the mathematical symmetry which it applied brought order to the realm of the hadrons. The infinite set of hadrons—the baryons and mesons—could be classified and the pattern of their properties illuminated. But like every major advance in physics, the success of the eightfold way raised new and deeper questions.

The main question was: Why did the eightfold way work? This question was especially puzzling in the 1960s because of the view physicists held at the time about the structure of hadrons—namely, that they didn't seem to *have* any definite structure. If a hadron was smashed apart, all that emerged was more hadrons created from the energy supplied by the smashing. The best explanation for this observation back in the 1960s was the bootstrap hypothesis, which asserted that all the hadrons were composed out of one another. To illustrate this idea, suppose that instead of an infinite number of hadrons there are only three called A, B, and C. First we ask: What is A made of? By smashing two A hadrons together we supply the energy that can create new particles, and in this case we learn that A is made up out of B and C. Next we do the same for B, which turns out to be made up out of A and C. Likewise, C is made out of A and B. The three particles are all mutually composite—they have lifted themselves into existence by "pulling on their own bootstraps." Many physicists in the 1960s were attracted to this bootstrap idea, which they now applied to an infinite set of hadrons, not just three, because it seemed to account for the fact that no new fundamental particles were seen in hadron collisions, just more of the same old hadrons. No hadron was more fundamental than any other; a "nuclear democracy" reigned in the microcosmos.

The difficulty with the bootstrap idea was that it gave no explanation for the eightfold way—the observed symmetry properties of the hadrons. The bootstrap did not seem the answer to the riddle of the hadrons; one had to look elsewhere.

The answer to the riddle of the hadrons first came in the mathematical imaginations of theoretical physicists. Murray Gell-Mann, and independently George Zweig, noticed that all the hadron families could be nicely accounted for if one imagined that the hadrons were built up out of more fundamental particles, which Gell-Mann called quarks. Using simple rules for combining quarks, all the infinite set of hadrons and the observed families were explained. The way to think of a hadron is as a little bag filled with a few elementary, pointlike quarks moving around inside. The observed new laws of charge conservation were just a consequence of the fact that different quarks were numerically conserved in hadron reactions—they were like atoms in a chemical reaction. The answer to the riddle of the hadrons is that the hadrons are quark "molecules." In 1969, Gell-Mann won a Nobel Prize for unraveling the symmetries of the hadrons.

In retrospect the hadrons seem like a trick played on physicists by the Producer of the 3-D movie. It was a trick which, once grasped, taught us not only the answer to the riddle of the hadrons but also about the unification of all the interactions of nature. But first—on to the quarks!

4. Quarks

—Three quarks for Muster Mark!
Sure he hasn't got much of a bark
And sure any he has it's all beside the mark.
—JAMES JOYCE, *Finnegans Wake*

HADRONS ARE MADE out of quarks: That is the answer to the riddle of the hadrons. But what are quarks? Quarks are point quantum particles similar to the electron and with the same spin of 1/2 as the electron. However, they have a fractional electric charge compared to the electron's one unit of charge, and unlike the electron, no one has ever seen a quark. Quarks came into modern physics not as a spectacular experimental discovery—a shout of "Eureka" from the laboratory—but as a mathematical trick of the theoretical physicists.

One day in 1963, Murray Gell-Mann was visiting Columbia University to deliver a lecture. Stimulated by questions and suggestions from Robert Serber, a theoretical physicist at Columbia, Gell-Mann came to the idea of a substructure to the hadrons which he called "quarks" after a line in Joyce's *Finnegans Wake*. "Quark" is a German word for a curd of cheese. Completely independently, another American physicist, George Zwieg, who was visiting CERN, the large international European nuclear laboratory near Geneva, came to the same idea, but he called the substructures "aces." Zwieg wanted to publish his article in an American physics journal, but the European leadership of CERN, intent on fostering a sense of independence from American physics, had a policy that research done at CERN had to appear in European journals. Zwieg's article was never published, and the term "aces" never caught on.

The basic idea of quarks is that all the hadrons can be built out of three quarks, called the "up" quark, "down" quark, and "strange" quark, and their three anti-quark partners (i.e., anti-matter versions of the quarks with opposite electric charge). "Up," "down," and "strange" are flavors of quarks—a curious use of the word "flavor." At one point physicists whimsically referred to the three quarks as "chocolate," "vanilla," "strawberry" instead of "up," "down," and "strange" and hence the use of "flavor." The ice-cream terminology never became popular, but the use of "flavor" as a generic label distinguishing the three quarks did. And it seems to be here to stay.

Physicists find it convenient to give letter names to particles, and the up, down, and strange quarks are denoted u, d, and s respectively and the anti-quarks $\bar{u}$, $\bar{d}$, and $\bar{s}$. A few properties of these quarks are listed in the quark table. A way to think of the quarks is as little point particles that bind together with strong forces to make up the hadrons—they are like a Tinker Toy set, the parts of which can be combined according to special construction rules to build up hadrons.

The rules for building the hadrons out of the quarks are very simple. The baryons, which we recall are one great subdivision of the hadrons with spin 1/2, 3/2, and so on, are combinations of three quarks,

$$qqq$$

where q can stand for either u, d, or s. Antibaryons are made of three antiquarks,

$$\bar{q}\bar{q}\bar{q}.$$

The other great subdivision of the hadrons, the mesons, which are hadrons with spin 0, 1, 2, and so on, are combinations of a quark and an antiquark,

$$\bar{q}q.$$

With these rules the hadrons are combinations of quarks that have integer electric charge 0, ± 1, $\pm$ 2, just the electric charges allowed for hadrons. The last rule is that you can only combine quarks together so that the total electric charge is an integer. That's it. With the u, d, and s quarks and these rules you can build up all the hadrons, and just the hadrons.

One obvious question is: How can you build the infinite set of hadrons out of just three quarks? There don't seem to be enough quarks to do this infinite job. According to the quark model, the quarks can bind together in lots of different configurations as

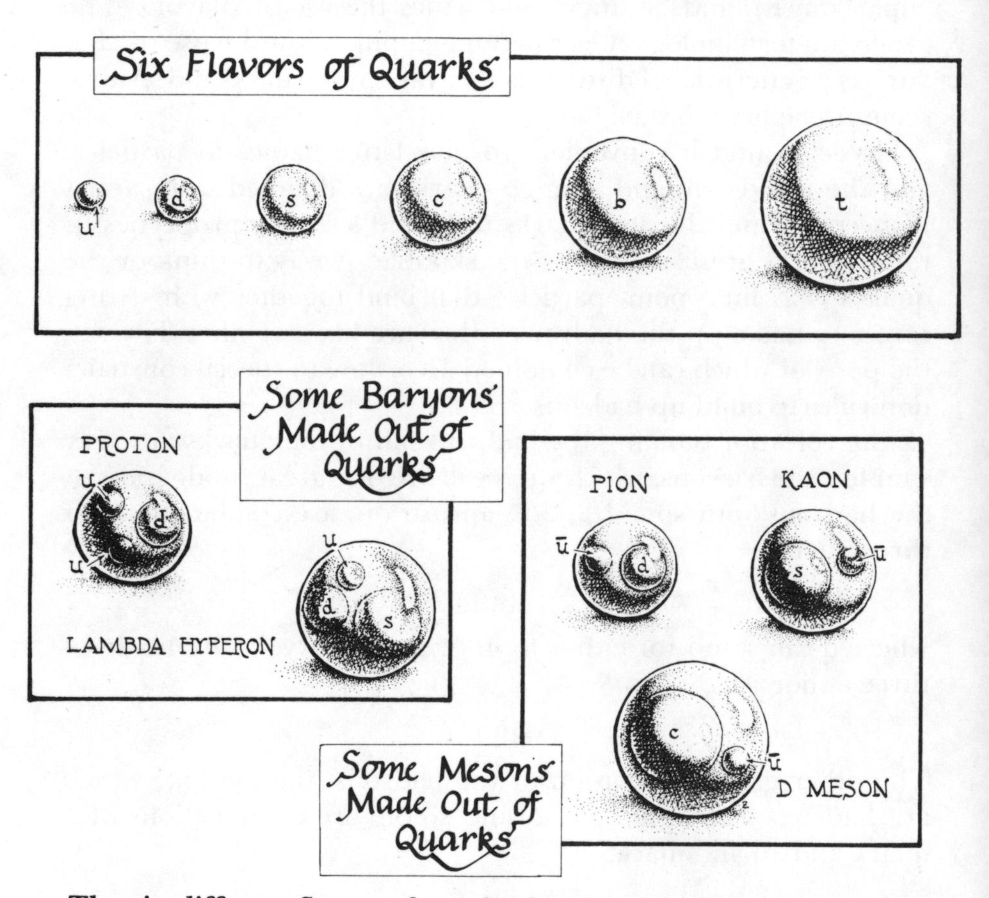

The six different flavors of quarks that theoretical physicists believe exist (five have been detected) and some baryons and mesons built out of these quarks. The flavors are up, down, strange, charm, bottom and top. The quarks have never been detected as free particles, only bound up inside the observed hadrons.

they orbit about each other in a hadron. Just like electrons in an atom, the quarks inside of a hadron have lots of orbits. For example, the quark and antiquark inside of a meson can orbit around a common center with one unit of orbital momentum, or two units, or three units, and so on, and each of these different configurations has different energy and corresponds to a different meson. Because there are an infinite number of orbits of the quark-antiquark pair, there is an infinite set of different mesons. In practice, physicists study only the lowest-energy orbital configurations that correspond to the most easily observed hadrons in the laboratory. The higher-energy orbits correspond to the hadron resonances that decay rapidly into lower-mass hadrons.

The conceptual beauty of the quark model is that it automatically explains the eightfold-way classification scheme for the hadrons that was previously known. Using the rules for combining quarks, the hadrons belong to families of 1, 8, 10 . . . members each—just those of the eightfold way. It would be as if the parts of our Tinker Toy set—the quarks—could be put together only into special combinations—the hadrons—that neatly fell into the family groups.

For example, the proton (p) and neutron (n)—the first hadrons to be found in the nucleus—are made out of u and d quarks according to

$$p \sim uud$$

$$n \sim udd$$

It is easy to check by adding the electric charges of the quarks that the proton has charge 1 and the neutron has charge 0. The proton and neutron, according to the eightfold way, are just two members of a family of eight hadrons. The rest of this octet comprising the strange friends of the proton and neutron are gotten by replacing one of the u and d quarks in the proton and neutron with the s, or strange, quark. The quark content of the resulting family of eight hadrons is shown in the diagram.

The only difference between the u, d, and s quarks so far as the

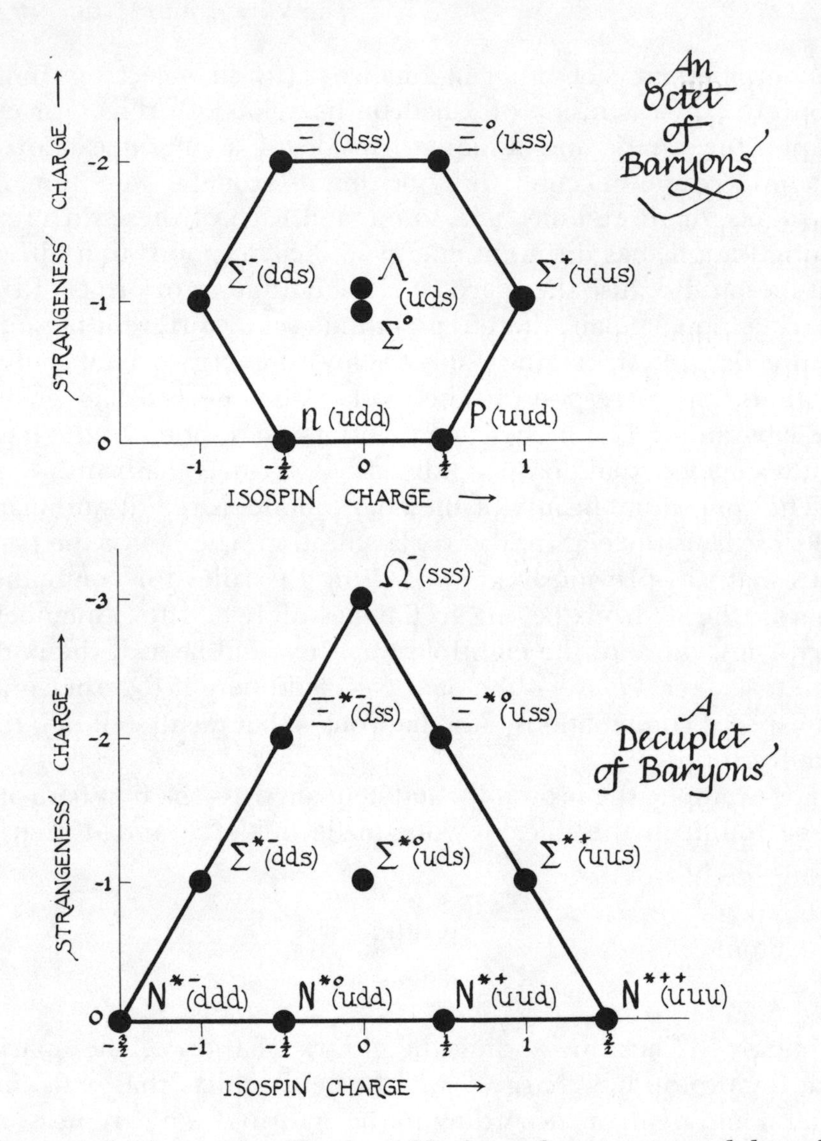

The eightfold way classification of hadrons for an octet and decuplet of baryons. Each baryon or meson detected in the laboratory fits into one such pattern with the strangeness charge plotted on the vertical axis and the isospin charge on the horizontal axis. Next to the symbol for a baryon is its quark content. With the advent of quarks the reason for these patterns became clear.

quark binding interactions is concerned is their mass—the u and d quarks are very light, while the strange quark is about fifty times more massive. No one has the slightest idea why these quark masses differ—it is an unsolved problem. But the fact that the mass of the strange quark is so much heavier accounts for the fact that strange hadrons, which must contain at least one strange quark, are heavier than nonstrange hadrons.

We can now appreciate the great simplification that the quark model represented—the problem of an infinite set of hadrons was reduced to a problem in the dynamics and interactions of just three quarks. The model accounted for the eightfold way. Further, the conservation laws that were observed in hadron interactions and led to the discovery of the eightfold way now were understood as the conservation of the various quark flavors—the same amount of "up," "down," and "strange" flavors that entered a reaction had to leave. The quarks exchanged in hadron collisions were like atoms being exchanged by molecules.

But there was a vexing problem: Quarks had never been seen. Quarks seem to exist only when they are bound together in the form of hadrons. Why? Perhaps free quarks do exist but have enormous mass and cannot be created in existing laboratories. Physicists have looked for quarks in laboratory experiments, cosmic rays, and other places, but without success. If physicists try to break a hadron apart into its quarks—like splitting a molecule into atoms—they do not get quarks, just more hadrons. The quark model was a mathematical fiction that somehow worked. How could physicists intellectually accept that fiction?

Physicists have a strong positivist streak, which is manifested by their never introducing any concept into physics unless it can be empirically verified in a direct way. Ernst Mach, an influential physicist at the turn of the century, never accepted atoms because he never saw one. Eventually physicists devised direct tests for the existence of atoms which previously were just convenient fictions for describing the behavior of gases. But what would Mach have thought of quarks? Most physicists today believe that quarks will never be seen—that they are permanently trapped in the hadrons. Although many physicists may be positivists, they are creative pragmatists first. The best physicists never let preconceptions influence the activity of their imagination—the imagination is to be guided by what works. And in this case, the quark model worked.

One can debate the existential status of quarks for a long time. The debate could end by the discovery of a free quark—a particle with fractional electric charge—and that would be the discovery of the century. New kinds of matter, never before seen, could then be created using quarks. A new chemistry with quarks instead of electrons and industries utilizing it would emerge. But I doubt that free quarks ever will be observed, because they have been sought for so long but not found. At any rate, the debate on the real existence of quarks as far as hadron structure was concerned ended around 1968. Like many issues of physics it was decided by experiment.

Shortly before 1968, a new instrument for the exploration of the structure of matter became available—the two-mile-long linear electron accelerator (SLAC) built in the hills behind Stanford University. A series of experiments carried out at Stanford by a local group and visitors from MIT convinced physicists that quarks existed inside of hadrons. A beam of electrons from the accelerator would scatter from a proton target by exchanging a single very-high-energy photon which could probe the electric charge inside the proton and measure its distribution. What was revealed was that the proton's electric charge was concentrated in pointlike structures. The quarks in the proton were like raisins in a pudding. In effect, physicists had looked inside the proton and seen the quarks.

These experiments done at SLAC to reveal the internal structure of the proton resembled the experiment done by Rutherford over fifty years earlier to determine the structure of the atom. Both experiments consisted of scattering one particle off another. Rutherford scattered alpha particles from gold atoms, and the SLAC experiment scattered electrons from protons. Rutherford did his experiment on top of a table, while the accelerator at

Stanford is two miles long—the difference in size of the equipment reflecting the factor of one billion in the energy required to "see" inside the atom versus that required to "see" inside the proton. But the basic experiment is the same.

After Rutherford announced his determination of the features of atomic structure, theoretical physicists like Niels Bohr turned to making models that would account for it. Likewise, after the SLAC experiment and confirming experiments from other laboratories, theoretical physicists turned to making models of the hadrons based on the idea of permanently confined quarks. Just as Bohr in making his model did not have a fundamental theory—the new quantum theory wasn't invented until later—theoretical physicists trying to account for the motion of the quarks inside the hadrons did not have a fundamental theory either. But that did not prevent them from making theoretical models of hadrons that could be compared with experiments and thus test the assumptions that went into the model.

One of the most successful models of the quarks in hadrons was the "bag model" proposed by Kenneth Johnson and his collaborators at MIT: Hadrons could be visualized as a tiny bag with the quarks inside. The bag or hadron can be thought of as an air bubble in a liquid, and within the bubble are the quarks, trapped permanently inside. Imagine a hadron collision in this model in which two hadrons or steam bubbles approach each other and then hit. For a short period of time the two bubbles actually overlap and are like a single bubble. During this time the quark passengers can jump from one side to another. Not only can the bubbles exchange quarks, but new quark-antiquark pairs can be created inside the overlapping bubbles from the energy of the collision. After the collision, the single bubble fragments into two or more bubbles, each representing a hadron with its quark passengers. The wonderful feature of this bag model is that you can calculate lots of details about hadrons, their collisions, and decays. The agreement with experiment is remarkable and strongly indicates that the idea of permanently trapped quarks is a workable one.

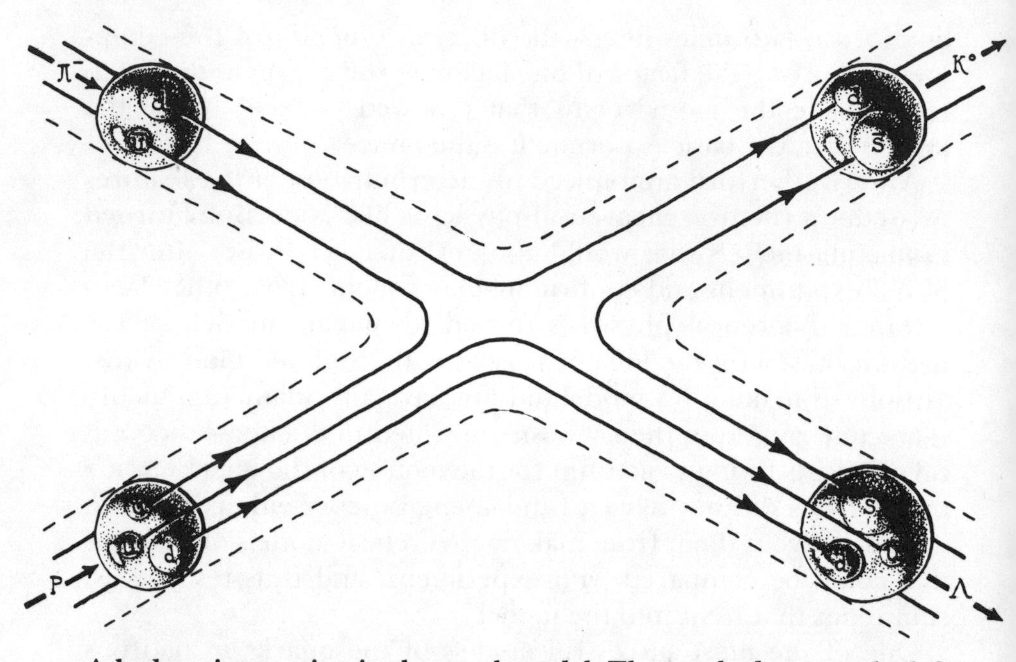

A hadron interaction in the quark model. The two hadrons on the left (a negatively charged pion and a positively charged proton) collide and transform into the hadrons on the right (a neutral kaon and a lamba). Quarks inside a hadron can annihilate corresponding anti-quarks or jump from hadron to hadron. New quarks can be created from the energy of the collision as are the s and $\bar{s}$ quarks in this collision.

What happens if we now try to pull the quarks apart from one another inside the bag? The bag begins to elongate and stretches out between the quarks that we are trying to pull apart. In fact, the hadron in this state looks less like a bubble and more like a string joining the quarks. This configuration describes what is called the "string model" of hadrons—the quarks are joined by some sort of glue that stretches into a string. We may imagine the string is like a rubber band. As we pull on it, the force remains constant. It would take an infinite amount of energy to pull two quarks apart if they were connected by such a string. Since an infinite amount of energy is not available, quarks cannot be separated.

But long before one can supply infinite energy, something else happens. The energy that you have supplied pulling on the string can turn into matter in the form of a quark-antiquark pair as shown in the figure. The quark-antiquark pair appears out of the vacuum because the energy exists to create them. The string breaks and becomes two strings or two hadrons. You never succeed in liberating the quarks. Instead, hadrons are created, just as observed in the laboratory.

The bag model and string model of the hadrons successfully account for many observed features of the hadrons. They give us an intuitive picture of hadron structure. But they are not fundamental quantum theories. Such a theory of the dynamics of quarks exists today and is called quantum chromodynamics. We

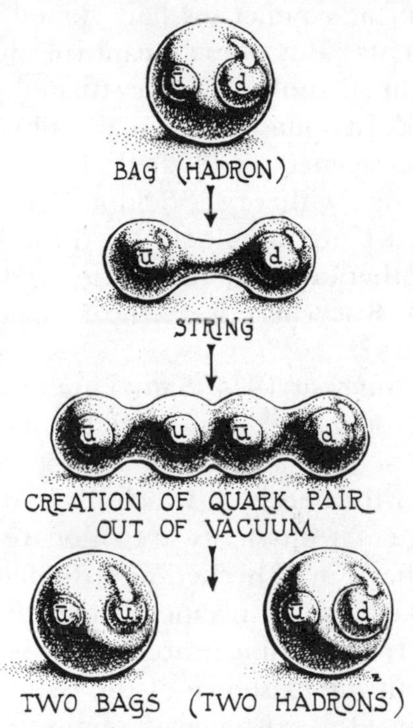

An unsuccessful attempt to liberate quarks from a hadron. The bag as it is stretched becomes a string. The energy used to separate the quarks is converted into creating a quark-antiquark pair. The result is two hadrons, not free quarks.

will describe it in a later chapter. The current hope of theoretical physicists is that the bag or string model with the quark confinement property can be derived from quantum chromodynamics just as Bohr's atomic model was derived from quantum mechanics. So far that goal has eluded us, but there is every indication that we are indeed on the right track.

Although the quark model with just three flavors of quarks accounted for the observed hadrons, some theoretical physicists as far back as the 1960s could not resist speculating that yet more quarks waited to be discovered. Most physicists paid little attention to these speculations, because there was no compelling reason for extra quarks. Around 1973 this situation changed dramatically. New theories of the interactions of quarks based on elegant mathematical symmetries had gained the confidence of theoretical physicists. But these beautiful theories would not agree with experiment unless one postulated the existence of a new fourth quark. In spite of this, many physicists still thought the fourth quark seemed farfetched. But the advocates of a fourth quark, led by the theorist Sheldon Glashow and his collaborators at Harvard University, held steadfast. Glashow had even dubbed the hypothetical new quark the "charmed" quark and insisted it existed. But where were the hadrons built out of the new charmed quark?

Around the summer of 1974, Sam Ting and his experimental collaborators at Brookhaven National Laboratory had noticed an unusual bump in some scattering data. Not many people heard about this result or the energy at which the bump was seen: It was a secret. Experimental physicists are often very secretive about their data until they have checked and double-checked it—they don't want to make an announcement which they later must withdraw. A physicist from the linear accelerator in Stanford, California, bet Ting about the existence of the bump to draw him out. But Ting bet against the existence of a bump, in spite of his data, perhaps to give the impression he had no result. In November of that same year an experimental group working at Stanford with electron-positron colliding beams did a careful search over different energies of the beams. At a particular energy an enormous number of particles was being created by the annihilating electrons and positrons, which showed up as a bump in the data plot. The Stanford experimentalists, led by Burton Richter, announced the discovery of a new particle which was implied by their bump. At the same time, Ting at Brookhaven immediately released his results. What the two teams of experimentalists had independently discovered was a new meson, a hadron built out of a charmed quark and an antiquark. Soon confirmation of the discovery came from a German laboratory. Shortly after this meson was discovered, yet another meson built out of the charmed quark c and antiquark $\bar{c}$ was discovered at Stanford. The collection of these mesons is called charmonium.

Physicists often refer to these surprising experimental discoveries as the November Revolution of 1974. Charm is just another flavor to be added to our list of up, down, and strange. As far as anyone has been able to determine, the charmed quark is similar to the previous three quarks except that it is simply far more massive. That is why its discovery had to await accelerators with the energy capable of creating it.

If the new hadrons discovered at Brookhaven and Stanford were really built out of a new charmed quark and an antiquark according to $\bar{c}c$, then there should exist even more new hadrons which are built up out of a charmed quark and another one of the old quarks u, d, or s. For example, there should exist new mesons built out of quarks according to $\bar{d}c$, $\bar{u}c$, $\bar{s}c$, which puts a charmed quark together with one of the previously known three quarks. This requirement really put the quark model to the test, and some of these mesons were in fact discovered in June 1976 at Stanford. A rich set of new hadrons can be built using the new building block, the charmed quark, a fact now confirmed by experiment. It is safe to say that there are no outstanding disagreements between quark model theory and experiment. The charmed quark seems to be just like the others except for its higher mass.

Now we have four quark flavors, u, d, s, and c. Why not more?

The great quark hunt was on, and it is probably not finished yet. In 1978 at the Fermi National Laboratory near Chicago, a group led by Leon Lederman, now the director of the lab, found another high-mass meson they dubbed the Υ, upsilon. It is undoubtedly a bound state of an even more massive quark, the "bottom" quark b (sometimes called "beauty"), according to the construction $\bar{b}b$. If this new quark is a more massive version of the other four quarks, it should combine with them also to form new hadrons. A "bare bottom" state, presumably the combination $\bar{u}b$, is being actively searched for at several laboratories.

Theoretical physicists think that there is at least one more flavored quark, the "top" quark t (sometimes called "truth"), waiting to be discovered at yet higher mass. It has been searched for but not yet found, and may exist at higher energies than we can currently achieve. How many quark flavors are there? No one knows for sure—that story will be for the future. But this much is certain: The problem of the hadrons has been reduced to the problem of the interacting quarks.

Quark Table

Name	Symbol	Approximate Mass in Units of the Electron's Mass	Electric Charge in Units of Proton Charge
UP	u	2	⅔
DOWN	d	6	−⅓
STRANGE	s	200	−⅓
CHARM	c	3000	⅔
BOTTOM	b	9000	−⅓
TOP	t	?	⅔

Today there are more quarks than there were hadrons in 1950, and the quark table is still growing. Many physicists feel uncomfortable with the large number of quarks, but no alternative has been devised. Are quarks the end of the road? Or are quarks themselves made out of more fundamental objects? Since quarks are apparently permanently confined inside hadrons, does it even make sense to talk about their having parts? As Sheldon Glashow remarks, "If this interpretation of *quark confinement* is correct, it suggests an ingenious way to terminate the apparently infinite regression of finer structure in matter. Atoms can be analyzed into electrons and nuclei, nuclei into protons and neutrons, and protons and neutrons into quarks, but the theory of quark confinement suggests the series stops here. It is difficult to imagine how a particle could have internal structure if the particle cannot even be created." All present evidence supports this view that quarks are a "rock bottom" to matter, but no physicist I know would be willing to bet much on that.

So far we have voyaged deep into the atomic nucleus, into the sea of hadrons to discover the quarks out of which they were all made. We now see that nuclear physics is the physics of trapped quarks. But there is another piece of the puzzle already seen in the atom—the electron. Where does the electron fit into the scheme of things? The electron belongs to another set of particles called leptons, which at first seem to have nothing to do with hadrons and their constituent quarks. They are another part of the cast in the 3-D movie, and next we turn to them.

5. Leptons

Who ordered that?
—I. I. RABI

A FRIEND OF mine, a theoretical physicist of Norwegian ancestry, used to sail the blue-white waters of Cape Cod in a small, swift sailboat, stirring up memories of his Viking past. He named his boat the *Lepton,* a Greek word denoting "light" or "swift." I recall sailing with him while the two of us puzzled about other leptons —a set of fundamental quanta which include the electron and the neutrino. Physicists knew about leptons and their properties for a long time, but no one knew how they fit into the plot of the cosmic 3-D movie. They seemed gratuitous—actors that no one needs. Only in the last several years have physicists seen how the leptons and quarks fit together into a unified theory of the fundamental quanta. But before going into that, let's first ask: What are the leptons?

Recall that in our voyage into matter as we reached the atom, we journeyed on into the atomic nucleus made of protons and neutrons. These two particles turned out to be but the first two out of an infinite number of hadrons. At the bottom of the riddle of the hadrons lay the quarks—just a few pointlike particles that bound themselves together with strong forces into an infinity of different configurations, the hadrons. The quarks are as far as the voyage into the nucleus has gone. But atoms had *two* primary components—the nucleus and the swarm of electrons surrounding it. What about the electron? Where does it fit in? The electron, physicists know today, is the first member of a new class of particles all with the same spin of 1/2 which go by the name leptons. The other leptons are the elusive neutrino, the muon, and the tauon, particles we will be describing here.

Why do physicists bother to classify the leptons separately from other particles like the hadrons and the quarks out of which they are made? Hadrons interact with each other very strongly, reflecting the strong forces that bind the quarks inside of them. By contrast, leptons have relatively weak interactions and thus make up a tidy corner of the world of the quanta. Physicists, realizing this, put the leptons into their own class.

Unlike quarks, which in many ways they resemble, leptons really can exist in the free state. The electron, for example, is bound to the nucleus of the atom by weak electromagnetic forces and is easily liberated. Physicists have made beams of liberated electrons, neutrinos, and muons. The electron gun in the base of a television picture tube fires a modulated beam of electrons at the screen to produce a picture. Leptons, like the electron, really exist out there in the world. Let us imagine opening the playbill for the cosmic 3-D movie to the part where the leptonic actors are described.

THE ELECTRON

The most conspicuous and mobile of the elementary quanta is the electron, already identified as a particle back in 1897. It is easily freed from its bonds to nuclei, and it has the lightest mass of all electrically charged quanta—a true, swift lepton. Electronic technology is a consequence of human mastery of the electron. The first use of electricity—moving electrons—was rather crude. All that was done was to move lots of electrons in the form of electric currents to light bulbs and power motors and all the other simple electrical devices. A more delicate management of electronic motion became possible with the development of the vacuum tube and then transistors. Scientists have learned to play with smaller and smaller amounts of electrons. Harnessing of the electron is the foremost example of how contact with the invisible quantum world has transformed our civilization. Because of the specific properties of electrons, which no one asked for or designed, electronic telecommunication and mass media became possible. The computer and microprocessor are transforming our civilization. The material laws of the invisible quantum world became the basis for new devices and opened a new sphere for human exploitation.

The modern theory of the electron begins with an investigation by Paul Dirac, one of the founders of the new quantum theory. Dirac knew that two separate sets of ideas—the new quantum theory and the relativity theory—were both correct. But the problem was to combine these two sets of ideas and devise a quantum theory which also obeyed the principle of relativity. Dirac focused his attention on the electron—a known quantum particle whose wave properties had already been confirmed. Could he find a mathematical description of the electron's wave which was consistent with Einstein's relativity theory? Finally Dirac mathematically deduced an equation—the Dirac equation—that the electron wave had to obey. This equation had profound implications.

First it predicted the observed properties of electrons as they moved in electric and magnetic fields, which, until then, eluded a deep theoretical understanding. These predictions are what convinced physicists that Dirac's theory was right. But the most startling prediction of Dirac's electron equation was the existence of a new kind of matter never before encountered—antimatter. How did this discovery come about?

Dirac's equation actually had two solutions; one solution described the electron, the other solution described a new particle with positive electric charge—opposite that of the electron. At first Dirac thought that the particle corresponding to this new solution had to be the proton, which at the time was the only positively charged particle known. Later it became clear that the new particle predicted by Dirac's equation had to have exactly the same mass as the electron and thus could not be the proton, whose mass was nearly 2,000 times that of the electron. Dirac's equation really predicted a new kind of electron—the antielectron or positron. Soon it was realized by physicists that the Dirac equation, in implying the existence of the antielectron, was offering but one specific example of the general consequence of combining quantum theory with the principle of relativity—namely, the existence of antimatter.

Antimatter is identical to ordinary matter except that the sign of all electric charges of the particles making up the antimatter are reversed. For every particle that can exist, the laws of physics imply that an antiparticle can also exist. Antiprotons could exist and so could antihydrogen atoms made up out of an antiproton and a positron. Whole antimatter worlds could exist—perhaps distant galaxies were made out of antimatter. Large pieces of antimatter do not exist in our world because matter and antimatter, if brought together, annihilate each other in a spectacular explosion. Antimatter seems bizarre—like another world behind the looking glass—but it was an unambiguous prediction of Dirac's theory, and if the theory was right it had to exist. And indeed it did. The antielectron was first discovered by Carl Anderson, a Caltech physicist, in cosmic rays raining down on the earth. Later, antimatter versions of the proton and other particles were found; physicists have even recently made a complete antideuteron, an atomic nucleus consisting of an antiproton and an antineutron.

Previous to Dirac's idea of antimatter, physicists could think of the quantum particles as immutable—the same number of particles that entered a reaction had to leave. But the discovery of antimatter profoundly changed that, as Heisenberg commented:

> I believe that the discovery of particles and antiparticles by Dirac has changed our whole outlook on atomic physics. . . . As soon as one knows that one can create pairs, then one has to consider an elementary particle as a compound system; because virtually it could be this particle plus a pair of particles plus two pair and so on, and so all of a sudden the whole idea of elementary particles has changed. Up to that time I think every physicist had thought of the elementary particles along the lines of the philosophy of Democritus, namely by considering these elementary particles as unchangeable units which are just given in nature and are always the same thing, they never change, they never can be transmuted into anything else. They

> are not dynamical systems, they just exist in themselves. After Dirac's discovery everything looked different, because one could ask, why should a proton not sometimes be a proton plus a pair of electron and positron and so on? . . . Thereby the problem of dividing matter had come into a different light.

As these remarks indicate, with the advent of antimatter, the notion of the conservation of the number of particles came to an end and the bootstrap hypothesis became feasible. Particles like the electron could be created and annihilated; they could transmute into one another.

The electron is the best understood of all the elementary quanta. It seems to be an absolutely stable particle. The reason for this is rather interesting. If we accept that electric charge is absolutely conserved—and most physicists today believe that it is—then since the electron is the lightest charged particle it cannot decay into any lighter particles because there is none that can carry away its electric charge. The electric charge is like a hand-me-down which, once passed to the youngest sibling, has no place further to go.

The modern theory of electrons interacting with light is called quantum electrodynamics and was invented just after the Second World War. This theory is one of the triumphs of modern theoretical physics, and has made predictions about the interactions of electrons which are verified by the most precise experiments ever carried out. Quantum electrodynamics incorporates Dirac's equation for the electron and indicates that the electron is a true point particle without further structure. Unlike protons and neutrons, which were made out of quarks, the electron seems to be at the end of the voyage as far as physicists can tell today. What compounds the puzzle of the electron is that it is not alone.

THE MUON

The second lepton, the muon, was discovered in 1937. Muons are the major component of cosmic radiation at the surface of the earth. At this moment lots of muons are flying around you and through your body. If we could be outfitted with special glasses that enabled us to see the muons around us, their tracks—provided they would last a minute—would form a dense thicket of nearly vertical lines around and through us. This invisible cosmic radiation can actually be revealed by detectors like spark chambers—devices which reveal the track of an electrically charged particle like the muon by a line of sparks between metal plates. But what is a muon?

The muon, as best as anyone can tell, is the same as an electron except that its mass is about 200 times greater. The muon is a fat electron. It is also described by the Dirac equation, it is negatively charged (the antimuon has positive charge), and its interactions are specified to great accuracy by quantum electrodynamics. Like all leptons it has no strong interactions.

After the muon and some of its properties were discovered, I. I. Rabi, the physicist, asked, "Who ordered that?" Rabi was expressing the feeling that no one needed or expected the muon. No one knows the answer to Rabi's question, but today we should generalize his question to include all quanta. No one has the slightest idea why leptons and quarks are in the cosmic 3-D movie.

THE NEUTRINOS

In the 1930s, physicists studying the radioactive decays of nuclei came across something that was very distressing. By precise measurements they found that there was more energy before the disintegration of the nucleus than after—a violation of the sacred law of mass-energy conservation. The theoretical physicist Wolfgang Pauli came to the rescue by suggesting that a new elusive particle was carrying off the undetected energy. At the time Pauli made this suggestion it seemed like a cheat to explain away the energy problem by postulating a new particle that was nearly impossible to detect. But eventually these particles were directly detected, and today physicists have made radiation beams of them. Unlike the electron and muon, these particles have no electric charge. Fermi named them "little neutral ones"—neutrinos.

Neutrinos are truly elusive leptons. They are lighter in mass than the electron (in fact, it is not clear if they have any mass) and have only extremely weak interactions with the rest of matter. They are often produced in the decay remnants of other particles. For example, the muon decays into an electron, a neutrino, and an antineutrino. Because it has only very weak interactions it is hard to stop a neutrino once it gets produced—it takes about eight solid *light-years* of lead to stop half the neutrinos emitted in a typical nuclear decay. They move like "greased lightning" through matter.

Amazingly, physicists can now produce and control beams of neutrinos such as the ones at Fermi National Accelerator Laboratory near Chicago and CERN, the European accelerator near Geneva. These intense, high-energy beams of neutrinos interact from time to time inside of enormous detectors. In spite of the small probability of such events actually happening, important experiments were performed using neutrino beams. Neutrinos, because they are so penetrating, can probe deep inside the structure of protons and neutrons, and much was learned about the quarks inside these particles. The unused neutrino beam, and that is most of it, after it leaves the detector just flies safely off through West Chicago. You can stand in a neutrino beam for days without a single event taking place inside your body.

One amusing proposal is to use a neutrino beam as a communication link. Theoretically, a neutrino beam could be shot straight through the earth and would emerge on the other side where a huge detector would pick up the weak signal. While in principle this could work, it would be very expensive compared to conventional methods. Another proposal, made tongue-in-cheek, is for a neutrino bomb, a pacifist's favorite weapon. Such a bomb, which could easily be as expensive as a conventional nuclear weapon, would explode with a whimper and flood the target area with a high flux of neutrinos. After terrifying everyone, the neutrinos would fly harmlessly through everything.

Physicists were surprised to find that there are two neutrinos, one of which is associated with the electron, the other with the muon. They are called the electron neutrino and the muon neutrino. But there was a further twist in the neutrino story: neutrinos are left-handed. Most fundamental quanta come in equal mixtures of right- and left-handed versions, but not neutrinos.

Some of the ordinary objects around us possess a "handedness" —they are either right-handed or left-handed. Gloves and shoes are an example—there is no way that you can turn a left-handed glove into a right-handed one (we will sew up the end of the glove so it cannot be turned inside out). Chemical molecules also can exhibit handedness, like the right-handed double helix of DNA that twists around like a spiral staircase to the right. Although DNA is right-handed, there is no fundamental law of physics or chemistry that says that life could not have developed using left-handed DNA. Life just has to use one or the other and on this planet the right-handed choice was made.

Neutrinos, if they are massless and like the photon always move at the speed of light, can also be either right- or left-handed. How can something so small and elusive as a neutrino have a handedness like a glove or DNA molecule? Neutrinos, like all leptons, have a spin of 1/2, and we can imagine them to be small spinning tops with their axis of rotation pointing in the direction of their motion. The spin rotation can be either clockwise or counterclockwise about the direction of forward motion, and these two possibilities correspond to right- and left-handed neutrinos respectively. If neutrinos always move at the speed of light, then we can never catch up to one and change its direction of relative motion and hence its handedness—a left-handed neutrino must remain one, and likewise for the right-handed variety.

Now we can appreciate why the neutrinos are so odd—there are no right-handed neutrinos. The fundamental laws of physics seem to forbid their existence—we couldn't make them if we tried. It would be as if there were a physical law that forbade the manufacture of right-handed gloves.

The fact that only left-handed neutrinos exist manifests a violation of parity conservation—the law that states that if a particle exists its mirror image (which changes right to left and left to right) can also exist. The mirror image of a left-handed neutrino is a right-handed neutrino and that beast simply does not exist. There is no way to make them. Two Chinese-American physicists, Chen Ning Yang and Tsung Dao Lee, developed in detail the suggestion that parity conservation was violated and proposed an experiment to test their hypothesis. When Pauli heard that experiments were underway to test for possible parity violation he remarked, "I do not believe that God is a weak lefthander." But the experiment of Chien Shung Wu and her collaborators at Columbia University showed that Pauli was wrong; God *is* a weak left-hander. Yang and Lee received a Nobel Prize for their novel work.

Physicists have always been fascinated by neutrinos, the lightest of all leptons. Recent experimental research has been focusing on the question of whether or not neutrinos actually possess a tiny mass or are massless. Speculations on the part of theoretical physicists suggest that they might have a small mass; if they do, it would have profound implications for cosmology, the study of the entire universe. If neutrinos had masses even a fraction of that of the electron they would then provide most of the mass of the universe. The elusive, invisible neutrino would be the dominant feature of the universe!

If you make a fist, there are thousands of neutrinos flying through it right now, because the entire universe is filled with neutrinos. In spite of their enormous numbers, they do not contribute much of anything to the total mass of the universe if they are massless. But if they have a mass then it is estimated that they would account for 90 percent of all the mass of the universe—an invisible mass, because no one can actually see this neutrino "background radiation." The other 10 percent of the mass of the universe—the minor part—is the visible matter in forms of stars and galaxies. Neutrinos could thus account for the "missing mass" of the universe—the amount required to halt the expansion of the universe and cause it, finally, to contract. Neutrinos could be the glue that holds the universe together. Whether neutrinos actually have a tiny mass is difficult to determine, but delicate experiments are now under way to settle this important question.

For a long time physicists thought that the electron, the muon, and their associated two neutrinos were the only leptons. But in 1977 there was a surprise.

THE TAU

The discovery of the tau came slowly and quietly. As early as 1976, physicists at a ring of colliding, counterrotating electrons and positrons near Stanford were seeing peculiar effects. The experimental leader of the group, Martin Perl, with cautious persistence suggested that these effects could be due to a new lepton, although other explanations could not be ruled out. In 1977–1978 with confirming evidence from a similar experimental facility at Hamburg, Germany, it was clear that there was a new lepton with a huge mass, 3,500 times the electron mass. Like the electron and muon, the tau presumably has a chargeless, left-handed neutrino associated with it, although there is little direct evidence for this. The tau, because it is so massive, can decay into lots of other lighter particles plus its associated neutrino. The only difference between the electron, the muon, and the tau seems to be their masses. If the muon was a fat electron, the tau was a fat muon.

One can hear Rabi's question—"Who ordered that?"—ringing in one's ears. Rabi's question expresses the puzzle of the leptons. Who needs any of the leptons beyond the electron and its neutrino? We need only the electron to build atoms. Yet the muon and the tau are just as fundamental as the electron. The remarkable feature of all the leptons is that they have never revealed any interior structures. They appear to be pure point particles even at the highest energies. This suggests they are truly elementary and not composite particles—a rock bottom to the levels of matter, a feature which compounds the puzzle of the leptons.

Physicists do not know the answer to Rabi's question or why the 3-D movie has all these extra characters. Theoretical physicists

Lepton Table

Name	Symbol	Mass in Units of Electron Mass	Electric Charge in Units of Proton Charge
ELECTRON	e^-	1	-1
ELECTRON NEUTRINO	ν_e	LESS THAN 0.00012	0
MUON	μ^-	207	-1
MUON NEUTRINO	ν_μ	LESS THAN 1.1	0
TAUON	τ^-	3491	-1
TAU NEUTRINO	ν_τ	LESS THAN 500	0

can fit these leptons into theories and make predictions about their interactions; but they do not know why they exist, why they have the masses they do, or any further explanation. The discovery of the tau seems to be the end of the lepton story for now. But many physicists would wager that there are heavier leptons waiting to be discovered once machines with the energy capable of creating them are built.

What we have seen of the cast of characters of the 3-D movie are the quarks and the leptons, both of which seem to be a "rock bottom" to matter. Leptons exist as real particles while quarks are trapped inside the hadrons. But how do the quarks and leptons interact with each other? We have said nothing about that. The interactions among these quarks and lepton actors in our 3-D movie are mediated by yet another set of quanta—the gluons. They are the final cast of characters in our 3-D movie.

6. Gluons

We call these quanta gluons, *and say that besides quarks there must be gluons to hold the quarks together.*
—RICHARD FEYNMAN

THE QUARKS AND leptons are the primary actors in the cosmic 3-D movie, and like real actors, these particles interact among themselves. Understanding the properties of these interactions took physicists most of this century; but today they understand them rather well. Using high-energy accelerators which probe matter, physicists found that the complexity of the world disappears at tiny distances, and the interaction between quantum particles becomes simple and falls into symmetrical patterns. A simple picture of reality emerged with the realization that the complicated interactions among quarks and leptons are in fact mediated by a definite set of quantum particles called gluons. As the name implies, gluons cause quantum particles to stick to one another—gluons are the glue that holds the world together.

Quarks, leptons, and gluons and their organization are all there is in the universe—the ultimate material, the final stuff from which all the complexity of existence emerges. They are the most distant shore on the voyage into matter to which physicists have traveled today. If there are further places to go, physicists have not surmised them yet. But what are the gluons and how did physicists come to understand their role in quantum interactions?

Physicists are always seeking patterns in the world which imply an underlying simplicity that is amenable to human comprehension. They are like detectives for whom patterns are the clues to reality. But what pattern exists in the complex interactions of matter with matter? No simple pattern of interactions is apparent if we look at the macroscopic world; but by examining matter on the microscopic level of the quanta—the atomic and subatomic levels—physicists have discovered that there are only four fundamental quantum interactions. In order of their increasing strength they are: the gravitational interaction; the weak interaction responsible for radioactivity; the electromagnetic interaction; and the strong quark-binding interaction. Each of these four interactions has an associated gluon, and the "stickiness" of the gluon is a measure of the strength of the interaction.

A remarkable feature of the quantum interactions of gluons—discovered only in this last decade—is that the strength of their interaction depends on the energy of the interacting particles. Quarks and leptons interacting at the relatively low energies available at present-day laboratories experience the four distinct interactions we gave above: the gravitational, weak, electromagnetic, and strong interaction. But the exciting discovery is that at much, much higher energy the strengths of the four interactions—the stickiness of the gluons—might all become equal and the distinctions between them vanish. The four interactions might really be the manifestation of but one universal interaction! This possibility is the basis for a unified field theory which has long been the dream of physicists.

In the last decade physicists have realized the dream of unifying the different interactions. Theories now exist which unify the electromagnetic, weak, and strong interaction into one interaction. Such unified field theories are the result of years of effort, and while theoretical physicists may argue over the details, the principle of unification is firmly established. We will examine these efforts in the coming chapters, but first we will describe how physicists think about the four primary interactions among the quanta.

A good place to begin thinking about quantum interactions is the simplest atom, hydrogen, in which a single electron is bound by the electric field to a proton, an example of the electromagnetic interaction. The old-fashioned way of thinking of the hydrogen atom is that there are two particles, the proton and electron, bound together by an electric field. A new way of thinking about the hydrogen atom which came out of the quantum theory is that the two quanta—the proton and electron—are exchanging a third quantum—the photon. Really there are not particles and fields; there are only quanta. We might think of the electron and proton as two tennis players hitting a ball—the photon—back and forth between them. This exchange of the ball binds the two players together—the photon acts as a kind of glue that holds the two components of the hydrogen atom together. The photon is the first example of a class of particles that physicists call gluons.

In this illustration we see the central idea of the modern concept of interactions—that interactions are mediated by quanta themselves. Each of the four fundamental interactions has associated quanta called gluons. The gluon associated with the electromagnetic interaction is the photon; the gluon of the gravity interaction is the graviton. Weak gluons mediate weak interactions, and colored gluons provide the quark binding forces. Particles such as quarks and leptons, by exchanging these gluons as two tennis players exchange a ball, can interact. And like a tennis ball, gluons can sometimes fly off on their own and may be directly detected as quantum particles. Let us now examine each of these four interactions in detail.

1. THE GRAVITATIONAL INTERACTION

Physicists describe the gravitational interaction as "long-range," meaning that it can extend out over macroscopic distances. The moon is bound to the earth by the gravitational interaction. Because gravity is long-range and stretches out over macroscopic distances its effects are evident in our surroundings. For this reason it was historically the first of the four interactions to be discovered. Yet from the standpoint of quantum particles it is the weakest of the fundamental interactions. The gravitational attraction of a single proton for an electron is over a billion billion billion (10^{27}) times smaller than the electromagnetic force. Only if there are large concentrations of matter, as in a planet or star, do the gravitational effects of all the particles add up and take on an observable role. The extreme weakness of gravity becomes apparent if we realize that the weight of our bodies is due to the entire mass of the earth interacting with it. As far as individual elementary particles are concerned we can ignore gravity because it is so very weak.

It is always interesting to see what the world would look like if we shut off one of the primary interactions. I like to imagine that in some distant place a demon sits at a huge control panel and operates this universe we live in. The demon can, by turning knobs and dials, adjust the fundamental physical constants which determine how our universe behaves. This demon can be viewed as either malicious or playful, but his job is to adjust the fundamental physical constants to make the universe as interesting and as livable as possible. Later his handiwork gets judged by a supreme demon who, should he be displeased, condemns the first demon to live in his own creation. This condition is designed to generate some human sympathy for the demon's job.

One of the main controls the demon has at his command changes the strength of the various interactions—it alters the stickiness of the gluons. Obviously, he will experiment with this control. If he turned off gravity, we would all fly off the earth and the earth and other planets would fly off their orbits around the sun. The sun, stars and large planets would cease to exist, because it is gravity that holds them together. The universe could consist of relatively small lumps of rocks like the asteroids held together by the chemical binding forces that hold rocks together. The effect of shutting off gravity would be very dramatic on the macroscopic scale but make little difference for the microcosmic world of quanta. Decreasing or shutting off gravity makes for an unlivable universe. But suppose our demon friend instead decides to increase the strength of gravity beyond its value today. Then stars and planets might collapse under their own weight into black holes—also an unattractive option. Probably our demon, if he is at all wise, will keep gravity at about its present strength.

According to the ideas of modern quantum field theory, which embrace the theoretical physicist's view of reality, every field like the gravity field has an associated quantum particle. For the gravity field these quanta are called gravitons; they are the gluons that bind large masses like stars together. Instead of thinking of the gravity field as some kind of force field that extends between the earth and moon, modern physicists see that such a gravity field is "quantized" into countless gravitons. Really the earth and the moon are exchanging gravitons, and these exchanges make up what we perceive as the gravitational field between these bodies. This is an unfamiliar but completely correct way of viewing the effect of gravity and other fields.

Although most physicists accept the idea that the gravity field is really quantized, it is unlikely that the graviton—the quantum of gravity—will ever be directly detected. In order to observe quantum particles like the graviton and not just their associated fields, one must look at the level of quantum interactions, and graviton interactions are simply too weak ever to be seen. Strictly speaking, if a graviton should hit a proton the proton would recoil. But this recoil is so tiny we can never detect it. Gravity is the weakling of quantum particle interactions.

2. ELECTROMAGNETIC INTERACTIONS

Like gravity, the electromagnetic force, seen in the form of electric and magnetic fields, is also long-range. But here the similarity to gravity ends. The electromagnetic interaction between charged particles is billions of billions times stronger than gravity. Unlike gravity, which has as its source masses which are always positive quantities, the source of electric and magnetic fields is moving electrically charged particles and these can be either positively or negatively charged. Hence the electromagnetic force can be either attractive (between oppositely charged particles) or repulsive (between like-charged particles), unlike gravity, which is always attractive (repulsive gravity or antigravity just doesn't seem to be allowed by our current theory of gravity). These are just some of the differences between the electromagnetic interaction and gravity.

The most interesting effects of the electric properties of matter are down at the levels of atoms. The reason for this is that most large quantities of matter are electrically uncharged and hence do not have electromagnetic interactions. However, individual particles in the atom, such as electrons, have associated electric fields which hold them in their orbits about the nucleus and are responsible, in part, for the chemical interactions of atoms. Almost all the properties of ordinary matter can be understood in terms of the quantum and electromagnetic properties of atoms. This understanding includes atomic physics, chemistry, condensed-matter physics, plasma physics—essentially all of physics except nuclear physics and cosmology, which require an understanding of the strong, weak, and gravitational forces. Because of its wealth of experimental implications the electromagnetic interaction has become the best understood of all the four interactions.

What would happen if our friend the demon shut off the electromagnetic interaction by turning a dial that reduces electric charges to zero? The most dramatic effect is that atoms would cease to exist and there would be no matter in the forms that we see about us. Atomic nuclei could now be very large, because there could be no electric repulsion between the like-charged protons, a factor which previously limited the size of nuclei. There would be proton stars like neutron stars—essentially gigantic nuclei. All the interactions between matter, except for gravity, would be very short-range. Life as we know it based on chemistry could not exist, and this is probably a good reason for the demon not to shut off the electromagnetic interaction.

The gluon associated with the electromagnetic interaction is the photon—the particle of light postulated by Einstein in his 1905 article on the photoelectric effect. At the time he postulated the photon few physicists believed it existed. But finally in 1923 experiments were done in which recoiling electrons hit by photons could be detected, and this convinced most physicists of the reality of the photon. The photon was the first—and so far the only—gluon which has been directly confirmed experimentally.

The most precise experiments ever devised are measurements of the electromagnetic interactions of photons and electrons. The modern theory that so successfully accounts for these experiments is called quantum electrodynamics, and it was invented in the 1920s by Werner Heisenberg, Wolfgang Pauli, and Paul Dirac. The photon was described by the quantized electromagnetic field and the electron by the quantized electron field. In the late 1940s, after many struggles with the mathematics, the final version of quantum electrodynamics was completed by Richard Feynman, Julian Schwinger, and Sin-itiro Tomonaga, an accomplishment for which they were awarded a Nobel Prize.

Quantum electrodynamics was the first practical example of what physicists call a relativistic quantum field theory. It was "relativistic" because it incorporated the principle of Einstein's special relativity theory; it was "quantum" because it embraced the ideas of the new quantum mechanics; and it was a "field theory" because the primary objects of its investigation were fields like the electric and magnetic field. Quantum electrodynamics incorporated the photon as the gluon of the electromagnetic field. It was enormously successful, so successful that it became the paradigm for future efforts to devise mathematical descriptions of the world of quanta. There is little question that it was the success of the photon concept and quantum electrodynamics that encouraged physicists to promote the view that all interactions were mediated by gluons. This viewpoint again met with success when physicists turned to examining the weak interaction.

3. THE WEAK INTERACTION

The weak interaction is responsible for the disintegration of many of the quantum particles encountered in the laboratory; in particular it accounts for radioactivity—the disintegration of atomic nuclei. In spite of the fact that there exist a very large number of quantum particles, only a small number of them, the electron, photon, proton, and neutrino, are observed to be stable —if left to themselves they do not disintegrate. Other particles, such as muons, neutrons, and other hadrons, will disintegrate rather rapidly into the stable ones. These particle decay processes provide an important clue as to the properties of the particles. Physicists have identified a special weak interaction as responsible for the decay. Unlike the electromagnetic and gravitational interactions, which can have long-range effects that we can see around us, the weak interaction is extremely short-range—its effects can be studied only by a careful examination of the quantum world. For this reason it took physicists a long time to understand the mysterious weak interaction.

Historically, humanity's first encounter with the weak force was the strange glow of radium salts seen in the late nineteenth century. As physicists investigated this, it became clear that the glow could not have a chemical origin—the energy release was far too great for such an explanation. Eventually the origin of the glow was established as due to particles emitted by the atomic nucleus. Physicists discovered that the nuclei of some atoms were unstable and as they disintegrated the emitted particles were detected as radioactivity. The study of the weak interaction took physicists into the nuclei and eventually into the subnuclear world of hadrons.

For decades there was much confusion in the theoretical and experimental account of these weak interactions that caused hadrons to decay. But today after much agony, struggle, and finally triumph, a theory exists which accounts for the experimentally observed properties of the weak interaction. The cornerstone of the theory is the assumption that, like the gravitational and electromagnetic interactions, the weak interaction is also mediated by gluons—"weak gluons." The weak gluons, unlike the graviton and photon, are extremely massive, so massive that no existing accelerator has the energy to create them. But larger accelerators are now being built that will have the energy to produce weak gluons, and most physicists anticipate that these very heavy gluons will be seen in the 1980s and will open up a new method for directly studying the weak interaction. Should the weak gluon be discovered it will be a major triumph for our current theories of the weak interaction.

How do the weak gluons cause particles like hadrons to disintegrate? The other gluons, the photon and the graviton we previously described, do not do that—why are the weak gluons so special? To understand how a hadron can disintegrate we have to remember what a hadron is made out of—the different-flavored quarks, up, down, strange, charmed, bottom, and top. What the weak gluons do is change the flavor of quarks, and that is how they can make hadrons disintegrate. For example, a strange quark in a hadron can be converted into an up or down quark by interacting with a weak gluon. This means that hadrons possessing a strange quark in them can change into hadrons with only up or down quarks in them—an example of a decay of strange hadrons. Likewise, charmed quarks can change into up and down quarks through the weak interaction. So this is the role of the weak gluons—they let strangeness and charm leak away, leaving only ordinary hadrons with, ultimately, only the proton as the remaining stable hadron. Weak gluons also interact with leptons and cause them to decay.

We can imagine what our friend the demon might find if he decided to shut off the weak interaction. Radioactivity would stop. But since radioactivity is hardly noticeable there would be no great immediate change in the world because of that. After only several million years the sun would cease shining because the weak interaction is used in its energy production. The big change would come from the fact that lots of exotic quanta like strange and charmed hadrons would now be completely stable. Because the weak interaction lets the "strangeness" and "charm" leak out of our world, it acts like a selective sink that drains away these exotic forms of matter should they come into existence. If we plug that drain by shutting off the weak interaction, then our world would be filled with strange and charmed particles just as it is filled with protons (which have no charm or strangeness). New kinds of matter—a new chemistry based on these now stable exotic particles—would be possible. The world would look extremely strange and far more complex than it does now. Different kinds of chemically based life might be possible. Conceivably, the demon just might decide to shut off the weak interactions, just to see what would result.

4. THE STRONG INTERACTION

The strongest interaction on our short list of four is the quark-binding interaction. The hadrons were made out of quarks—but what holds the quarks together? Why don't they just fly apart when hadrons collide? The answer that theoretical physicists have come up with is that the quarks are bound together by a new set of gluons which are so super-sticky that the quarks never can come unglued. The necessity for these new gluons became apparent in the same famous electron-scattering experiments done at Stanford that first saw the quarks inside the proton. As Richard Feynman remarked:

> . . . if we add up all the momenta of the quarks and antiquarks which we see in the electron and neutrino scattering experiments, the total does not account for the momentum of the proton but only for about half of it. This must mean that there are other parts in the proton that are electrically neutral and do not interact with neutrinos. Yes, and even in our model of three quarks we had to hold the quarks together somehow, so they could interact and exchange momenta. This may well be via some interaction field (analogous to the electric field which holds atoms together) and this field would carry momentum and would have quanta (analogous to the photons). We call these quanta *gluons,* and say that besides quarks there must be gluons to hold the quarks together. These gluons contribute the other half of the momentum of the proton.

It was clear that once again the idea that interactions were mediated by gluons was vindicated. Physicists turned their attention to understanding these new quark-binding gluons, and soon a new theory was born—quantum chromodynamics. This was a relativistic quantum field theory that gave a mathematical description of these strong gluons just as quantum electrodynamics gave a description of the photon.

The main idea of quantum chromodynamics is that each of the quarks we discussed previously—the up, down, strange, charmed, top, and bottom quarks—also comes in three "colors." Of course, quarks really are not colored any more than they have flavors—that is just a way of talking that helps us form a picture in our mind. The "color" of quarks refers to a new set of charges—if you like, labels—for the quarks. The new strong quark-binding gluons stick or couple to the color charges in the same way a photon couples to electric charge. According to quantum chromodynamics, there are eight "colored gluons" that provide the strong quark-binding forces.

These forces, mediated by the colored gluons, are supposed to be so strong that all quanta which possess a colored charge (so the colored gluons will stick to them) are permanently bound together. Consequently the quarks, since they have the colored charge, are permanently bound to each other by the color gluons. Even the colored gluons, because they have the color charge, also stick to themselves so tightly they never become unglued. However, there may be combinations of colored quarks and gluons for which the sum of all the colored charges add to zero just as the combination of the positive electric charge of an atomic nucleus cancels the negative charge of the orbiting electrons so the atom is electrically neutral. Such "color neutral" combinations of colored quarks correspond exactly to the observed hadrons. This set of ideas, embraced by the chromodynamic theory, has lots of experimental support.

Suppose our demon shut off the colored strong forces. This implies that the colored gluons would fly off and quarks would be liberated. There would be no hadrons, no protons, neutrons, pions—the whole hadron zoo would disappear into its constituent quarks flying about. Atoms would not exist, because the nucleus made out of protons and neutrons would not exist. Maybe quarks could form some other kind of matter. But the world would certainly be very different. A wise demon would probably not experiment too much with the quark-binding forces.

We have completed our tour of the four fundamental interactions. At first sight, gravity, electromagnetism, the weak, and the strong force have nothing to do with one another. But as physicists explored deeper into the structure of matter the distinction between the interactions became superficial. That is the major message of theoretical physics in the last decade.

Each interaction is mediated by quanta—the gluons. The gluon of gravity is the graviton, of electromagnetism the photon, of the weak interaction the weak gluons, and of the strong quark-binding force the colored gluons. Remarkably, each of these gluons is mathematically described by a relativistic quantum field theory, each of which is rather similar to the others. Ten years ago, anyone suggesting that the theory of the strong interaction would

Gluon Table

Name and Symbol	Couples to	Role in Quantum Interactions
GRAVITON	MASS	GRAVITY; COSMOLOGY; BINDS PLANETS TO THE SUN, AND STARS TO GALAXY
PHOTON γ	ELECTRIC CHARGE	ELECTROMAGNETIC WAVES; BINDS ELECTRONS TO NUCLEUS
WEAK GLUONS W^+, W^-, Z°	WEAK, FLAVOR CHARGES	RADIOACTIVE DECAY PROCESSES OF HADRONS AND LEPTONS
COLORED GLUONS	COLORED CHARGES	STRONG FORCE, BINDS QUARKS PERMANENTLY INSIDE HADRONS

resemble the electromagnetic interaction would have been laughed at. Today we know quantum chromodynamics and quantum electrodynamics are not so different.

It is held today that all the four interactions become unified as one universal interaction at ultrahigh energies. The reason we see such different interactions is that the familiar world we know has its physical processes occurring at low energy, and there the strengths of the interactions can be very different. The symmetry and simplicity of physics will be revealed only at ultrahigh energies. Unified field theories—and many have been constructed in the last several years—incorporate these features.

According to these unified theories, all the interactions we see in the present world are the asymmetrical remnant of a once perfectly symmetrical world. This symmetrical world is revealed only at very high energies, energies so high they will never be accomplished by human beings. The only time that such energies existed was in the first nanoseconds of the big bang which was the origin of the universe.

If we go back to the beginning of time, to the first moments of creation, the energy of the primordial fireball was so high that the four interactions were unified as one highly symmetrical interaction. As this fireball of swirling quarks, colored gluons, electrons, and photons expanded, the universe cooled and the perfect symmetry began to break. First gravity was distinguished from the other interactions, and then the strong, weak, and electromagnetic interactions became apparent as they froze out of the cooling universe manifesting symmetry breaking. Exotic quanta like charmed particles decayed away, and soon mostly protons, neutrons, electrons, photons, and neutrinos were all that was left. After still further cooling, atoms could form and condense into stars; galaxies developed and planets emerged. As the surface temperatures of some planets dropped, complex molecules began to form—the building blocks for life. Even in the evolution of life we see this process of symmetry breaking at work as organisms generally evolved from simple to more complex ones. Human societies, too, seem to become more complex as they develop. The universe from its very beginning to the present may be viewed as a hierarchy of successively broken symmetries—a transition from a simple perfect symmetry at the beginning of time to the complex patterns of broken symmetries we see today.

At those immense energies at the beginning of time, life could not exist. Although interactions were unified and perfectly symmetrical, it was a sterile world. The universe had to cool and the perfect symmetries break before the complex interactions that gave rise to life could exist. Our world manifests a broken or imperfect symmetry. But out of that imperfection arose the possibility of life.

7. Fields, Particles, and Reality

In its essentials, this point of view has survived to the present day, and forms the central dogma of quantum field theory: the essential reality is a set of fields *subject to the rules of special relativity and quantum mechanics; all else can be derived as a consequence of the quantum dynamics of those fields.*
—Steven Weinberg

In the last few chapters we have presented the cast of the 3-D movie: quarks, leptons, and gluons. These quantum objects are at the "rock bottom" of the material world—everything we know of can be made out of them. Now we turn to examining the script for the cosmic 3-D movie—the laws these quanta obey. How do physicists describe subatomic particles like electrons, photons, and protons? What is going on at those tiny distances?

The discovery of the levels of matter—molecules, atoms, nuclei, hadrons, and quarks—and the interactions of quanta raised fundamental questions about the very concept of matter, questions addressed by the theoretical physicists. What is the "stuff" out of which particles like electrons and photons are made? At the beginning of this century, although physicists knew about electrons and photons, they could not answer such questions. Today they can. A coherent, unified, and experimentally correct picture of material reality was invented for thinking about the world of quanta. How did these new ideas develop?

The first three decades of this century saw the creation of two great physical theories, Einstein's theory of special relativity and the new quantum theory. These theories, taken together, provide the conceptual framework of almost all of physics and are the basis of our ideas about material reality. Confronted by these two theories, physicists in the late 1920s asked: Was it possible to merge the relativity and the quantum theory into a single theory that embraced both sets of principles?

Accomplishing this merger was far more difficult than theoretical physicists had at first thought. After years of intellectual struggle the result of the merger was the invention of relativistic quantum field theory. A working example of such a theory was intact in the late 1940s. It was called quantum electrodynamics, and it dramatically changed the way that physicists thought about matter.

Relativistic quantum field theory implied that to understand atomic particles one has to go beyond the old idea of matter as a "material stuff" that can be known through our senses to descriptions of particles in terms of how they transform when subject to various interactions. It is how material objects respond when acted upon that tells us what they are.

We can't see subatomic particles. Suppose for the moment that we can't see tennis balls either but can observe a tennis match played with invisible balls. By observing the serves, swings, and motions of the players we could begin to estimate the mass and size of the object they are exchanging. We could conclude a lot about the ball even though we can't see it. Careful observations would lead us to deduce that the invisible ball could spin. We could tell it wasn't a golf ball or football by how hard the players hit the ball, how they moved. That is the way it is with subatomic particles—we can't see them either. But by examining how instruments respond to them we can establish their properties and the laws of their motion. Subatomic particles do not obey the laws of motion of tennis balls given by Newton's classical physics. They obey the weird laws of motion given by the quantum theory, and, as Richard Feynman commented, "they are all crazy in the same way."

The quantum theory describes the interaction of subatomic

particles through the field concept. At first it seems that a particle has nothing to do with a field like a magnetic field, but as we describe the field concept it should become clear that particle and field are complementary manifestations of the same thing.

Familiar examples of fields are the gravitational and magnetic fields, which are long-range and can extend over large distances. A field is invisible, but we can determine its effects. We can imagine that the mass of the earth is a source for the gravity field which pervades all of space and attracts material bodies to the earth. Similarly, the magnetic field of the earth or a bar magnet can be visualized as a force field pervading space. The effect of the magnetic field can be seen on a compass—a fact that fascinated the four-year-old Einstein.

In the old Newtonian theory of gravity, the actual existence of the gravity field was not required and it did not have any material reality. It was simply a useful mathematical fiction for describing the effect of gravity on particles of matter. One could describe gravity as well without it.

The field concept—that fields have a physical existence—came into its own in the nineteenth century. Michael Faraday, the English physicist who did extensive experiments on electricity and magnetism, especially emphasized the physical nature of the electric and magnetic fields. Electrically charged particles were viewed by him as points at which the field became infinitely large—the field, he argued, was the essential physical object, not the particle. Faraday's intuition about the physical nature of the field was finally realized in James Maxwell's electromagnetic theory of light in which light is a wave of oscillating electric and magnetic fields propagating in space. Electric and magnetic fields were not mathematical fictions in Maxwell's theory but could carry energy and momentum. Fields had a physical reality.

That light is a form of energy is clear to us simply by standing in the sun. The heat we feel was once energy in the form of light—electromagnetic fields—transmitted across interplanetary space from the sun. But light also carries momentum and exerts a "radiation pressure"—a small but observable push. We could set up large sails on spaceships in outer space that would catch the "light wind" from the sun to sail between the planets. Such a regatta might take place in the next century.

In spite of the advances of nineteenth-century physics, there were two dualisms that had to be overcome before the modern concept of matter could develop. First was the dualism of mass and energy, which were seen as distinct. This dualism was overcome by Einstein's relativity theory, which showed that mass and energy were convertible; mass was simply a form of bound energy. The second dualism was that of field and particle often referred to as the wave-particle duality. This dualism was overcome by the new quantum theory, in which fields and particles were no longer seen as distinct but as complementary.

Before the advent of quantum theory, physicists thought of particles and fields as distinct entities. For example, the electron and proton making up the hydrogen atom were understood as particles bound together by the electric field of the mutual attraction. Particles were considered to be immutable and eternal. Fields emanated from particles and were responsible for the forces between them. While this picture of particles and fields as distinct seemed adequate at the time, there was the troublesome puzzle of Einstein's photon—light as a particle. How could light be both an electromagnetic wave field as in Maxwell's theory and a particle as required in Einstein's explanation of the photoelectric effect? A great step toward resolving this question was de Broglie's suggestion that particles like the electron had associated wave fields. Particles could behave like wave fields and wave fields like particles. What was going on?

The dualism of fields and particles was overthrown in the late 1920s. With the development of the quantum theory fields took on a new significance. As Steven Weinberg remarked:

> In 1926, Born, Heisenberg, and Jordan turned their attention to the electromagnetic field in empty space and . . . were able to show that the energy of each mode of oscillation of an electromagnetic field is quantized. . . . Thus the application of quantum mechanics to the electromagnetic field had at last put

> Einstein's idea of the photon on a firm mathematical foundation. . . . However, the world was still conceived to be composed of two very different ingredients—particles and fields—which were both to be described in terms of quantum mechanics, but in very different ways. Material particles, like electrons and protons, were conceived to be eternal. . . . On the other hand, photons were supposed to be merely a manifestation of an underlying entity, the quantized electromagnetic field, and could be freely created and destroyed. It was not long before a way was found out of this distasteful dualism, toward a truly unified view of nature. The essential steps were taken in a 1928 paper of Jordan and Eugene Wigner, and then in a pair of long papers in 1929–30 by Heisenberg and Pauli. They showed that material particles could be understood as the quanta of various fields, in just the same way as the photon is the quantum of the electromagnetic field. There was supposed to be one field for each type of elementary particle. Thus, the inhabitants of the universe were conceived to be a set of fields—an electron field, a proton field, an electromagnetic field—and particles were reduced to mere epiphenomena. In its essentials, this point of view has survived to the present day, and forms the central dogma of quantum field theory: *the essential reality is a set of fields* subject to the rules of special relativity and quantum mechanics; all else is derived as a consequence of the quantum dynamics of those fields.

These ideas mark the beginning of relativistic quantum field theory—the merger of relativity and quantum theory. The microworld and indeed the whole world could be viewed as a vast arena of interacting fields. Previously, physicists imagined the world was divided into matter and energy. The matter resided in particles and the energy in fields that interacted with the particles, causing them to move. Now a unified view was established. The dualisms of energy and matter, particle and field, were dissolved, and everything could be seen to be interacting quantum fields. There isn't anything to material reality except the transformation and organization of field quanta—that is all there is. This marked the ultimate triumph of the field concept in the human attempt to comprehend reality.

According to the quantum theory, the intensity of the field at a point in space is interpreted as the statistical probability for finding its associated quanta—the particles. What is meant by "quantizing a field" is analyzing a field like an electromagnetic wave in terms of its associated quanta, the photons. The intensity of the electromagnetic field at a point in space gives us the odds for finding a photon there.

The notion that reality is a set of fields that give the probabilities for finding their associated quanta is the most important consequence of relativistic quantum field theory. It is the central concept for the picture of reality. Not only did the idea of matter disappear into the field concept, but the field specified the probability for finding quanta. God rolls the dice every time a quantum interaction takes place.

The world of interacting quantum fields is not easy to visualize. We can describe that world in terms of mathematics and make our concepts precise, but it is like trying to imagine objects in an infinite dimensional space—the visual imagination fails to produce an adequate picture. But we can get some feeling for quantum field theory from the following analogy, the infinite 3-D mattress.

Take an ordinary steel spring and imagine that it is floating in space. Attach identical springs to the ends of that spring and more springs to the ends of those until a grid lattice of steel springs is constructed that pervades all of three-dimensional space. This is the 3-D mattress. This entire lattice of springs represents a quantum field in our analogy. Let's suppose it is the electron field. If a single spring in the lattice is plucked, it will vibrate, and this vibration corresponds to the quantum, an electron associated with the field. Two springs far apart on the lattice could also be separately plucked and the resulting vibrations would correspond to two quanta, two electrons at those points.

We could imagine a second mattress made of a different kind of spring, maybe a heavier one, that is superimposed on the first lattice, and this mattress would represent a quark field. Its vibrations correspond to quark particles. So for each field there is a different mattress of springs pervading all of space, and the vibrations of a specific spring correspond to a particle at that point.

So far these different superimposed lattices of springs representing all the fields of nature are supposed not to touch each other. But now imagine that the various lattices of springs representing quarks and leptons can get linked together by another set of springs, which represent the gluons. The electron lattice is linked to the photon lattice and this to the quark lattice and so on. This space of connected spring lattices now represents an interacting quantum field theory.

If one of the springs of the electron field lattice has a vibration corresponding to an electron at some point, this vibration can be transferred to the photon field lattice. That now starts vibrating corresponding to photons in the neighborhood of the electron. The photon could also couple to the quark, and so on. All fields —lattices of different kinds of springs, one for each particle—can interact with each other via some third kind of field.

To press this analogy even further, we imagine that the springs are invisible. All that remains of the lattice of springs is its set of vibrations. Further, the individual springs should be made infinitely small so that even in a tiny region of space there are infinitely many springs. This super 3-D mattress of tiny invisible springs is rather close to what theoretical physicists describe as a quantum field. All that remains of the field is its potential vibrations at each point—the quanta manifested as various particles. These particles can move about in space and interact with one another. The underlying reality is the set of fields, but its manifestation is the particles. The universe is a great spawning ground and battlefield of the quanta, according to relativistic quantum field theory.

With this background we can now list the central dogmas of relativistic quantum field theory:

1. The essential material reality is a set of fields.
2. The fields obey the principles of special relativity and quantum theory.

3. The intensity of a field at a point gives the probability for finding its associated quanta—the fundamental particles that are observed by experimentalists.
4. The fields interact and imply interactions of their associated quanta. These interactions are mediated by quanta themselves.
5. There isn't anything else.

These points make up the conceptual framework of modern relativistic quantum physics. They give us the basic picture of reality. Within this framework physicists must try to account for all of physics. There are several fundamental questions that arise from these central dogmas.

First, what are the fundamental fields? Since each field has an associated quantum particle, we are just asking again what the fundamental particles are. This question has been answered by experimental physicists, who, in their voyage into matter, have found the quarks, leptons, and gluons. Right now these quanta seem to represent the most elementary levels of matter, and hence their fields are the most fundamental.

The second question which occupies many theoretical physicists is that once the fundamental fields have been identified, how does one make a mathematical model—a relativistic quantum field theory—that describes their interactions? Examples of such field theories of the known quanta are quantum electrodynamics and quantum chromodynamics.

People are always interested by the question of whether physicists invent these theories or discover them as Columbus discovered America. Are theories "out there" in the world waiting for some bold and clever person to find them? I don't think so—theories are inventions. I like to think of physical theories as programs for a computer which, starting with little data, calculates everything about the quantum interactions that we can compare to observations. The design of such a computer program can of course fail. But if the program works, then our invention shows us something about reality in the only way we can get at reality. From relativistic quantum field theories a new picture of material reality has emerged—not only in terms of its content, the fundamental quanta, but also the very concept of reality, the stage on which the cosmic 3-D movie is enacted.

Our aim in the coming chapters will be nothing less than to present the concept of material reality as understood by physicists today. A good place to begin our topic is with the physicist's concept of nothing at all—the vacuum.

8. Being and Nothingness

No point is more central than this, that empty space is not empty. It is the seat of the most violent physics.
—John A. Wheeler

"Nature," said Aristotle, "abhors a vacuum." He observed that whenever we try to remove all matter from a region of space the tendency is for matter to rush in and fill the void. Matter is everywhere. Our modern conception is just the opposite—matter is the exception in the universe. Most of the space between the stars is empty or nearly so, and even solid matter is mostly empty space with all the mass concentrated in the tiny atomic nuclei. Almost everything is vacuum.

However, the old idea of the vacuum—that it is empty space, nothingness—has also changed. After inventing relativistic quantum field theory in the 1930s and 1940s, physicists came to a new concept of the vacuum—it is not empty; it is a plenum. The vacuum, empty space, actually consists of particles and antiparticles being spontaneously created and annihilated. All the quanta that physicists have discovered or ever will discover are being created and destroyed in the Armageddon that is the vacuum. How can this possibly be?

Space looks empty only because this great creation and destruction of all the quanta takes place over such short times and distances. Over long distances the vacuum appears placid and smooth—like the ocean which appears quite smooth when we fly high above it in a jet plane. But at the surface of the ocean, close up to it in a small boat, the sea can be high and fluctuating with great waves. Similarly, the vacuum fluctuates with the creation and destruction of quanta if we look closely at it. Even when looking at the level of atoms these vacuum fluctuations of the quanta are extremely small, but observable. On the basis of measurements of atomic energy levels, physicists know the vacuum fluctuations are really there, and if they could look at even smaller distances the vacuum would be revealed as a churning sea of all the quanta. Instead of "Nature abhors a vacuum" the view of the new physics suggests, "The vacuum is all of physics." Everything that ever existed or can exist is already potentially there in the nothingness of space. Physicists came to this remarkable view of the vacuum by way of a deeper understanding of Heisenberg's uncertainty principle and the existence of antiparticles. Here is how it works.

A rigorous law of modern physics is the law of conservation of energy. We can imagine that this law is checked by an energy accountant who keeps track of the total energy in a physical interaction. In his ledger he records energy expenses and energy income, and the two columns have to balance exactly. That is the law of energy conservation.

The Heisenberg uncertainty relation comes into play if we now apply this law to the world of quantum interactions. The uncertainty relation implies that if we measure the energy of a quantum like the electron over a short but definite time interval then the degree of uncertainty in the measurement of that energy is inversely proportional to that time interval. Thus, for very short time intervals there can be great uncertainty in our knowledge of the energy of the quantum. This means that the energy accountant, for short periods of time, must make errors in the energy income and expense column, although the errors average out to zero over long time periods. The uncertainty relation implied a loophole in the argument that the law of energy conservation meant that quanta could not be created out of nothing. They can be created out of nothing for short time periods. The errors in the energy account are like the waves on the vacuum sea. In some places the waves are higher, in other places lower, but they average out to what we see from on high—a smooth sea. The random errors our energy accountant makes are just another manifestation of the statistical nature of reality and the dice-playing God. The vacuum randomly fluctuates between being and nothingness.

Since energy is uncertain for short time periods, a quantum could, in principle, come into existence in empty space and then quickly disappear. Such a quantum that goes in and then out of reality is called a virtual quantum. It could become a real quantum, an actual particle, only if it had sufficient energy to do so. These virtual quanta are like the errors the energy accountant is making. They have a virtual reality, but in the end they have to cancel out. If we could supply the needed energy to the vacuum from an external source, then the virtual particles in the vacuum could become real. It would be like telling the energy accountant that he had a real credit in the account and one of his income errors didn't have to be canceled by an expense error. This process of the creation of real from virtual quanta has actually been observed in the laboratory.

The virtual quanta in the vacuum should properly be thought of as pairs of particles consisting of a virtual particle and its antiparticle. A vacuum fluctuation consists of a particle and its antiparticle springing into virtual existence at a point in space and then immediately annihilating each other. The lower the mass of such pairs of particles, the more likely it is to create them out of the vacuum, since the energy error required to do so is small. Consequently the biggest waves on the vacuum sea are electron-positron (antielectron) pairs, since these are the lowest-mass particles. Smaller waves correspond to heavier particle-antiparticle pairs. But there are waves on the vacuum sea corresponding to every conceivable quantum, even those we have not yet discovered. All of physics—everything we hope to know—is waiting in the vacuum to be discovered.

Another way to visualize the vacuum is to imagine the 3-D mattress as an analogue of the quantum field as we described before. The springs of the mattress covered all of space and were infinitely small, and the vibration of a spring corresponded to a quantum particle. We might think of the vacuum as the lattice of springs with no spring vibrating—the absence of real particles.

However, because of the Heisenberg uncertainty principle we can never be sure that a spring has strictly no vibration. So the springs are allowed to vibrate below the level actually corresponding to real particles. These vibrations correspond to the virtual quanta —the waves on the ocean. If we were to supply real energy to those vibrations they could increase to the level of becoming real particles. The vacuum is filled with the vibrations of every possible quantum.

This remarkable concept of nothingness was a consequence of the new theoretical ideas supporting relativistic quantum field theory. Although the new idea of the vacuum was understood on a theoretical basis, the question of the experimental verification of this fantastic idea was open. How could physicists determine the effect of these virtual quanta in the vacuum?

Most of the space between an atomic nucleus and an orbiting electron can be thought of as empty, and this is where physicists found the new vacuum effects. Indeed, one effect of the creation and annihilation of virtual quanta were tiny shifts in the energy of the orbiting electron about the atomic nucleus. What happens is that the electric field that binds the electron in orbit about the nucleus can sometimes create an electron-positron pair out of the churning sea of virtual quanta in the vacuum. The pair then immediately annihilates. This effect, called vacuum polarization, will slightly change the orbit of the electron about the nucleus. One of these orbital changes of the electron in the hydrogen atom was measured with great precision by the experimentalist Willis Lamb. He used precision microwave techniques that were developed with radar during the Second World War. The interesting feature of Lamb's measurements was that they could be compared with theoretical calculations based on the quantum field theory called quantum electrodynamics. Failure to take this exotic effect of vacuum polarization into account would have led to a disagreement with Lamb's observations. Instead, the calculated value of the energy of the electron orbit accorded precisely with his measurements. The shadow world of the virtual quanta in the vacuum had a real effect.

Although there were many other experiments that checked with the vacuum polarization effects predicted by theorists, the most dramatic vindication of this new concept of the vacuum was the construction of the electron-positron colliding beam machines. These machines, which first became operational in the 1970s, collided a high-energy beam of electrons (matter) against an oppositely directed beam of positrons (antimatter). The collisions of matter with antimatter provided the necessary energy to bring the virtual pairs of particles fluctuating in vacuum into real existence.

These colliding beam machines, by supplying energy to the vacuum, actually probe the structure of the vacuum in terms of virtual particle-antiparticle pairs. Quark-antiquark pairs of particles pulled into existence from the vacuum can be created in this way. That is one way that the new quarks, like the charmed quark, were discovered. The charmed quark and antiquark pair was but a tiny wave on the vacuum ocean which, once physicists supplied just the right amount of energy, could be brought into existence in the form of a new hadron. Physicists anticipate that even more new forms of matter will be discovered using this technique of bringing the virtual vacuum quanta into existence. That will be the story for experimental physics in the 1980s—creation of matter by supplying colliding beam energy to the vacuum.

Once our minds accept the mutability of matter and the new idea of the vacuum, we can speculate on the origin of the biggest thing we know—the universe. Maybe the universe itself sprang into existence out of nothingness—a gigantic vacuum fluctuation which we know today as the big bang. Remarkably, the laws of modern physics allow for this possibility. Aristotle thought the universe had existed forever. But his careful reader, Thomas Aquinas, disagreed with him and instead thought the world was a *creatio ex nihilo*—a creation out of nothing. The entire universe could be a representation of nothingness—the vacuum.

It is ironic how physics turned out in this century. The nineteenth and early twentieth century was characterized by a materialist outlook which maintained a sharp distinction between what actually was in the world and what wasn't. Today that distinction still exists, but its meaning has altered. What doesn't exist, nothingness or the vacuum, is a kind of joke by the "eternal Maker of enigmas." Theoretical and experimental physicists are now studying nothing at all—the vacuum. But that nothingness contains all of being.

9. Identity and Difference

Tweedledum and Tweedledee
Agreed to have a battle
—LEWIS CARROLL,
Through the Looking Glass

THE INDUSTRIAL WORLD of mass manufacturing surrounds us with artifacts that appear to be identical. In the supermarket I am confronted with rows of identical cans of food, the products of a machine civilization. Automobile parts, functionally identical, are interchangeable. I am struck by the fact that the experience of an environment of identical things must be rather recent, because in the ancient world artifacts were made by hand and had differences. Even ancient coins, intended to be alike, were each different except to the most casual observer.

The tendency of the human mind is to look for differences. Identical things are confusing, even terrifying, reducing the sense of our own uniqueness. I find something dreadful in the black-and-white photographs that Diane Arbus has taken of twin children, usually dressed identically. The children never appear happy. What is dreadful is the implied threat to our own identity; these children are like cans in a supermarket, like the parts of automobiles.

Confronted with identical objects, many people feel uncomfortable. Perhaps there is a biological basis for this feeling, because human survival has sometimes depended on our capacity to discern small differences. If we look closely at "identical" objects then we will usually see small differences, a scratch or a nick. All macroscopic objects have discernible differences, and somehow this perception is comforting. But if we enter the microworld of molecules and atoms we also enter the world of absolute identity. There is no use in looking for differences, because two molecules or atoms in the same energy state are absolutely identical. Atoms have no scratches on them to tell them apart.

If we go one step further to the level of elementary particles like electrons, then there is no question of difference. Quantum particles have no internal structure that distinguishes them—two electrons are absolutely identical, and so are two photons. The truth is that the entire material universe, with all its variety, is entirely made up out of quantum particles which are completely identical. It was nature, not nineteenth-century manufacturers, that first utilized the principle of interchangeable parts. The fact that one electron is absolutely identical to the next has important physical consequences which we now will explore.

Our story begins with Leibniz, the philosopher who first formulated the principle of the identity of indiscernibles, which states that if there is no way to establish the difference between two objects then they are identical. It is a definition of identity, and one that most people consider obvious. Among other things it would imply that interchanging the positions of two identical objects does not change the physical state of those two objects. While interesting philosophically, this interchange "symmetry" of identical objects had no observable consequences in the old classical physics. But with the invention of the new quantum theory in 1926 the identity of indiscernibles took on a remarkable significance. Physicists realized that not only are the quantum particles such as electrons and photons absolutely identical but the consequence of this identity is the existence of a new kind of force between them. The identity of indiscernibles has, as a consequence, a force! Without these new forces, called exchange forces, chemistry and atoms as we know them could not exist—we would not be here. With the aid of the basic concepts of the quantum theory and some elementary mathematics we can show how the identity of particles implies the existence of these new exchange forces.

Imagine that there are two identical particles like two electrons or two photons sitting at two points in space, x_1 and x_2. According to the quantum theory, these two particles are completely described by a probability wave whose shape depends on the points x_1 and x_2—mathematically, the shape is a function of those two points. In particular, the shape of the probability wave that describes the particles will depend on $x = x_1 - x_2$, the distance between the two particles, and we will denote this wave shape by $\Psi(x)$.

Although the two particles are completely described by the probability wave shape $\Psi(x)$, we must remember that according to the quantum theory and Born's statistical interpretation the probability, $P(x)$, for finding the particles a distance x apart is *not* equal to the wave shape $\Psi(x)$ but to the intensity of the wave, $P(x) = [\Psi(x)]^2$, obtained by squaring the wave shape. Since it is probability that we can measure and observe, it is really the square or intensity of the wave shape that has physical meaning.

Using these ideas of the statistical interpretation of the wave shape, we can now apply the identity of indiscernibles. This principle asserts that if we interchange the two identical particles by interchanging their positions we should not be able to observe any difference. Interchanging the particles means interchanging their positions x_1 and x_2. Hence the distance separating the particles $x_1 - x_2 = x$ becomes $x_2 - x_1 = -x$, that is, it is changed to its negative.

The wave shape $\Psi(x)$ is a function of x, and so is its square, the probability $P(x) = [\Psi(x)]^2$. If the probability for finding two identical particles does not change if we interchange them, then we cannot observe any physical difference. So the principle of indiscernibles implies that the probability, $P(x)$, should not alter if we interchange the two particles. Mathematically this is expressed by

$$P(x) = P(-x)$$

What does this imply about the probability wave shape $\Psi(x)$? One possibility is that the wave shape $\Psi(x)$ is an even function of

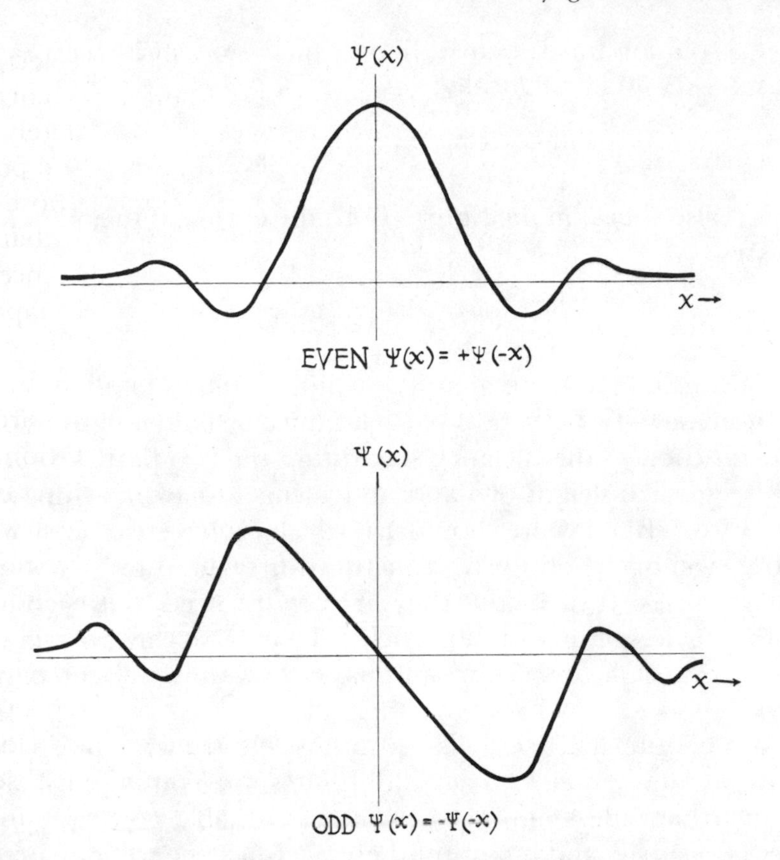

Even and odd wave shapes as a function of the separation of identical particles. Notice that for the even wave shape the probability (given by the square of the wave shape) for finding two particles on top of each other (x = 0) is greatest, while for the odd wave shape the probability for finding the particles on top of each other is zero.

the separation x—it does not change as $x \rightarrow -x$. Mathematically this is expressed by

$$\Psi(x) = +\Psi(-x) \qquad \text{(even)}$$

An example of such a wave shape is shown in the figure. There is, however, one other possibility. Since $(-1)^2 = +1$, we can also have a solution for the wave shape which is odd—it changes sign —as $x \to -x$. Mathematically:

$$\Psi(x) = -\Psi(-x) \qquad \text{(odd)}$$

This is also shown in the figure. Both the even and the odd choice satisfy

$$[\Psi(x)]^2 = [\Psi(-x)]^2$$

We conclude that the identity of indiscernibles implies that the wave shape of two identical particles must be either an even or an odd function of the distance separating the two particles. So far this seems like sleight of hand and seems to have nothing to do with forces. But now we should ask which of these two possibilities —the even or the odd wave function—is realized by actual identical particles. It turns out they are *both* realized, but each for a different class of quantum particles. To explain this, we will have to make a slight excursion and say a few words about particle spin.

All quantum particles, like photons, electrons, protons, neutrons, and even quarks, have a definite spin. You might imagine them to be little spinning tops. A remarkable consequence of special relativity and quantum theory—which we will not attempt to prove—is that the spin of these quantum particles is quantized. The spin cannot be arbitrary; it must take on discrete values. The spin of these particles in special units of spin then takes on the values

$$0, 1/2, 1, 3/2, 2, 5/2 \ldots$$

The amount of spin is either an integer 0, 1, 2 . . . or half-integer 1/2, 3/2, 5/2 . . . , a distinction which is of considerable importance in quantum theory. Particles with integer spin, like the photon with spin 1, and half-integer spin, like the electron with spin 1/2, behave very differently.

So we see there are *two* families of particles, those with integer spin 0, 1, 2 . . . and those with half-integer 1/2, 3/2, 5/2. . . . In our discussion of the interchange of a pair of particles we discovered that the identity of indiscernibles led to *two* choices for the wave function—even or odd. There is a celebrated theorem in quantum theory—the spin-statistics theorem—which we can now state: For integer spin particles one must always choose the even wave function, while for half-integer spin particles one must always choose the odd wave function. So *both* the even and the odd choice have physical importance—each applies to a different family of particles.

It is now a simple matter to see how the identity of indiscernibles can lead to forces between identical particles. Suppose we have two identical photons separated by a distance x. The photon has spin 1, so the wave shape $\Psi(x) = +\Psi(-x)$ that describes the two photons must be an even function of x, the distance between the two photons. Because it is an even function of x, it need not vanish at $x = 0$, corresponding to two identical photons sitting right on top of each other. This indicates that there is a definite probability for two photons to be on top of each other. If the probability is higher that they are on top of each other than separate, then it appears there is a "force" that causes them to attract. In fact there is no "real" force but only a higher probability that the photons are near rather than far apart. How can a probability be manifested as a force?

Recall our discussion about rolling a pair of dice? The probability was highest that a 7 would be rolled and lowest for a 2 or 12. It would appear that after many rolls as if the dice were "attracted" to the 7 more than the 2 and 12. This example is quite general—if an event has a high probability it seems as if there is an attractive "force" that makes it happen. Conversely, if the probability for an event is low it is as if there is a repulsive "force" that prevents it from happening. It is just probabilities, but it appears like a "force," what physicists call an exchange force. In the quantum theory such exchange forces take on a physical significance.

We have seen that the identity of indiscernibles and quantum theory imply an attractive force between photons. If we have a gas of lots of identical particles, which like the photon have integer spin, and if it is cooled down to very low temperature so that the motion of the particles gets slowed down, these attractive forces begin to dominate. A condensate—what is called the Bose-Einstein condensate—of these particles will form because of the attraction, and this has been experimentally observed. What seems like a sleight of hand with probabilities has reality.

But the implications are even more remarkable if we consider two half-integer spin particles like electrons. Their wave shape must be odd under the interchange of positions, $\Psi(x) = -\Psi(-x)$. Consequently, if we put two electrons on top of each other so x, the separation between them, is zero, $x = 0$, we have $\Psi(0) = -\Psi(0) = 0$. The wave shape must vanish at zero separation, since the only number that is equal to its negative is zero. We conclude that the probability of finding two electrons on top of each other is zero—just the opposite to what we found for photons. Two electrons, if you try to put them into the same place, experience what amounts to a repulsive force. The rule that two electrons cannot sit on top of each other is called the Pauli exclusion principle.

Pauli's exclusion principle explained the periodic table of the chemical elements on a theoretical basis. According to this principle, two identical electrons could not occupy the same place or the same orbit around the atomic nucleus. As one added more electrons in orbit about the nucleus, building up different chemical elements, each electron had to be put into a new state. When this was done using Pauli's exclusion principle, it was clear that the periodic table of chemical elements discovered by Mendeleev could be exactly accounted for: Quantum theory explained the chemical properties of the elements. For over a century, ever since the discovery of chemical elements, such an explanation was sought. Now it was at hand.

But quantum theory went beyond just establishing the theoretical basis for the periodic table. The way in which atoms combine to form molecules could also be finally understood. Several atoms making up a molecule can share electrons. In sharing electrons the "odd" property of electronic wave shape comes into play along with its associated exchange force, and these exchange forces bind the atoms together in particular molecular configurations. Understanding the rules for these exchange forces led to the theory of the molecular bond by Linus Pauling, among others. This discovery marks the beginning of quantum chemistry. Chemists and physicists could actually calculate the angles and distances between atoms in molecules and compare their calculations with the results of X-ray data which measured these same angles and distances.

The quantum theory and the new role of the identity of indiscernibles provided the fundamental basis for the understanding of all of chemistry. Of course, some organic molecules are gigantic and can twist and turn upon themselves in a very complicated way. Scientists do not understand all the foldings of these big molecules precisely. But that is because they are so complicated, not because there is a lack of understanding of the fundamental chemical forces. For those complicated molecules, computers are an essential tool to solve the puzzle of their structure from data. It is only a matter of time before the fundamental quantum theory will explain even the most complicated molecules.

The quantum theory has immeasurably deepened our understanding of the identity of indiscernibles and the problem of identity and difference. The absolute identity of quantum particles in the hand of a God that plays dice takes on a new significance, providing new forces between particles. These exchange forces are crucial for binding electrons to nuclei in the very special way that gives rise to the laws of chemistry. In fact, without the repulsive electronic exchange force—a consequence of sleight of hand of quantum probabilities—atoms would collapse and we would not be here. It is one of the more subtle tricks of the dice-playing God.

I have always been fascinated by an analogy between language and the world of quantum particles. Written English language is based on the alphabet, a set of twenty-six different letters plus some punctuation marks. By ordered arrangement those distinct letters can make words and sentences. Although one letter *a* is identical to another letter *a,* words and sentences can be different —a rich set of possibilities is open. Likewise, in our universe there are only a rather few fundamental building blocks: quarks, leptons, and gluons. These are the letters in the alphabet of nature. With this rather small alphabet, words are made—these are atoms. The words are strung together, with their own special grammar—the laws of quantum theory—to form sentences, which are molecules. Soon we have books, entire libraries, made out of molecular "sentences." The universe is like a library in which the words are atoms. Just look at what has been written with these hundred words! Our own bodies are books in that library, specified by the organization of molecules. The universe as a literature is, of course, a metaphor—both the universe and literature are organizations of identical, interchangeable objects; they are information systems.

We now see that the significance of absolute identity uncovered in quantum theory is that it explains the very forces that bind the world together. Out of identity came difference—the difference of every thing from every other thing in our world.

10. The Gauge Field Theory Revolution

Nature seems to take advantage of the simple mathematical representations of the symmetry laws. When one pauses to consider the elegance and the beautiful perfection of the mathematical reasoning involved and contrast it with the complex and far-reaching physical consequences, a deep sense of respect for the power of the symmetry laws never fails to develop.

—C. N. YANG, Nobel Prize lecture

ANYONE WHO HAS even the most superficial contact with physics is impressed by the inherent simplicity and beauty of the laws of nature. How can the laws of physics be simple if the world is so complicated? The answer to this question is one of the great discoveries of Newton. He saw that all the complications of the world lay in the specification of initial conditions—the positions and velocities of all particles at one instant of time. The laws of physics that described how the world changes from such initial conditions could be, and were, very simple. This insight, separating the complicated initial conditions from the simple laws of physics, survived to the present day.

Still, this does not explain why the laws of physics are in fact simple. Only our experience, gained over several centuries, supports our conviction in the simplicity of physical law, a conviction which recently has been powerfully reinforced by the success of relativistic quantum field theories of the fundamental quanta.

The new ideas brought forth by relativistic quantum field theory—antimatter, the new physics of the vacuum, identical particles and exchange forces—transformed the way physicists thought about reality. Theoretical physicists realized that the idea that material reality was a set of fields provided the key to understanding the fundamental interactions among quanta—the gravitational, the weak, the electromagnetic, and strong interaction. They hoped that the mathematics of field theory would give a detailed description of quantum particles the same way that centuries earlier Newton found that the mathematics of differential equations gave a description of ordinary, classical particles. Developing quantum field theory took physicists into new branches of mathematics such as infinite-dimensional Hilbert spaces, operator theory, and matrix algebra, once again vindicating the remarkable notion that the fundamental laws of nature are specified in terms of beautiful mathematics.

But what impressed theoretical physicists most of all is the profound role of symmetry, described by the mathematics of group theory, in elucidating the laws of the quantum interactions. As physicists came to understand the mathematical symmetries of field theory they discovered that these symmetries actually required the very interactions they had observed among the fields and their associated quanta. The remarkable insight that symmetry itself implies the existence of gluons that mediate the interactions among quarks and leptons is the subject of this chapter. It is an example of how the simple and beautiful ideas of symmetry lie at the very foundation of the complex quantum interactions, ideas which because they were simple and beautiful could be grasped and appreciated by the mind.

After the invention of relativistic quantum field theory in the early 1930s, some mathematical physicists focused on developing it as a new mathematical discipline for its own sake, while others applied field theory to reality—the 3-D movie of quarks, leptons, and gluons. These physicists sought specific theories to describe the quantum particles, theories that would enable them to perform calculations that could be compared against experimental observation. Guided by experiments, many of which were done at the great accelerator laboratories, these theoretical physicists eventually discovered the field theories such as quantum electrodynamics, which described the interaction of photons and electrons, and quantum chromodynamics, which described the quark-binding strong interaction. The intellectual work that went into constructing these field theories spans five decades and is still going on. But the consensus of physicists today is that field theory represents a major triumph in the quest to comprehend material reality. Physicists can now precisely describe in mathematical language the interactions of quarks, leptons, and gluons. Mathematical reason has enabled the human mind to comprehend the farthest shore on the voyage into matter. How did this amazing triumph of physics come about? What were the crucial ideas that led to the success of field theory?

Modern field theory had its birth in the attempt to unite quantum mechanics and relativity in a single theory. Theoretical physicists back in the 1930s first turned to this problem through the study of photons and electrons. The problem which stood like a mountain before them was to make a mathematically consistent and experimentally correct relativistic quantum field theory which described the *interaction* of these two quanta.

When physicists first attempted to calculate the interaction of photons and electrons they found that the numbers they obtained from their calculations were infinite, a nonsense answer. What could be going wrong? The reason these infinities arise can be seen from our image of a quantum field as a kind of 3-D mattress of springs. Recall that the springs were to be infinitely tiny and fill all of space. This means that no matter how small a region of space we examine there are infinite numbers of springs that can vibrate. These vibrations correspond to virtual quanta, and the fact that all these quanta exist even at the tiniest distances means that if we calculate the influence of such virtual quanta on the mass of the electron, we obtain an infinite answer for the mass—complete nonsense.

Some physicists thought that quantum field theory made no sense because of this problem with infinitely large quantities. But other physicists held firm to the belief that the basic ideas of field theory could be made to work. Eventually they managed to tame these infinities in a mathematical *tour de force* called the renormalization procedure. Here is how it works.

Suppose a man weighs 150 pounds on a bathroom scale. Then he has a good dinner and adds a couple of embarrassing pounds. But he decides to cheat by adjusting the bathroom scale so that it continues to read only 150 pounds. This cheating—or rescaling—is the renormalization procedure. If someone could actually put on an infinite amount of weight and then reset the scale by an infinite amount to give a finite weight, then one gets an idea of the amount of cheating or renormalization required in the calculational procedures of quantum field theories. Remarkably, for some field theories this cheating can be carried out in a mathematically consistent way, and these are called "renormalizable" quantum field theories.

The renormalization procedure showed theoretical physicists that some relativistic quantum field theories made sense because they were insensitive to the structure of matter at the very shortest distances. As we learned in the chapter on matter microscopes, shorter and shorter distances are probed by quantum particles of higher and higher energy and momentum, and hence the very shortest distances, corresponding to a very high momentum, are beyond the reach of experiment. Julian Schwinger, one of the inventors of the renormalization procedure, remarked, "Somehow, the process of renormalization has removed the unbalanced reference to very high momenta that the unrenormalized equations display. The renormalized equations only make use of physics that is reasonably well established." Schwinger's remarks emphasize that the importance of the renormalization procedure was that the interaction of photons with electrons at low momenta and energy, which physicists could study experimentally, could now be mathematically described without detailed knowledge of what was happening at very high energy. For reasons still quite unknown, nature has chosen renormalizable theories to describe the quantum interactions. The correspondence between mathematics that we can solve and the real world seems like a lucky break.

By the late 1940s, theoretical physicists had for the first time a field theory which obeyed both the principles of quantum theory and relativity theory. It was called quantum electrodynamics, and it was a remarkable accomplishment. Using the renormalization procedure, physicists could now confidently calculate the interactions of virtual photons with electrons and compare the results with experiments. The agreement was remarkable. As experimental physicists refined the measurements of the electromagnetic properties of the electron further and further, theoretical physicists calculated these same properties and matched the measurements number for number. In many ways this triumph of quantum electrodynamics was similar to the triumph of Newton's classical physics applied to the motions of the planets—incredibly precise astronomical observations were accounted for by mathematical calculations from theory. And just as the experimental success of Newton's theory convinced physicists of the correctness of classical physics, so too the experimental success of quantum electrodynamics convinced physicists of the correctness of field theory applied to photons and electrons. All those exotic virtual quanta and the new physics of the vacuum were not artifacts of a mathematical formalism, a product of our imagination, but were really required to rationally understand the properties of matter. Of all the modern quantum field theories, quantum electrodynamics is the exemplar of success.

Spurred on by this triumph of field theory, theoretical physicists next turned to the problem of finding a similar theory for the strongly interacting quanta—the hadrons. At that time in the 1950s and 1960s physicists did not know that hadrons were made out of quarks. They just went ahead and assumed each hadron had a fundamental field associated with it. But when they applied field theory to the strong interactions of hadrons they utterly failed to make progress. Why?

For one thing, there were a large number of hadrons, and they all interacted with one another in a complicated way. The interactions were so complicated that it was never possible to consider the simple case of just two hadrons interacting; they all got into the act. A second problem was that the interaction between hadrons was about one hundred times stronger than that between electrons and photons, so strong it was difficult to understand it mathematically.

Because of these difficulties, many theoretical physicists—perhaps even a majority—thought at one time of abandoning field theory and replacing it with a completely different approach called the S-matrix (S for "scattering") theory. The advocates of S-matrix theory felt that the trouble with field theory was that it introduced objects into physics—the fundamental fields—which were beyond the reach of experiment. By contrast, the S-matrix theory was invented to employ only experimentally observable quantities and attempted to relate one set of measurements on hadron interactions to other, similar sets of measurements. The hope of the S-matrix theorists was that they would never have to introduce any concept into their theory which was not related to an experimentally observable quantity. Field theorists did not care if they introduced such concepts; as long as they could calculate observed quantities they would be satisfied.

In many ways the S-matrix theory drew its inspiration from Ernst Mach's philosophy of physics—that physics is a science of measurable objects and events and that physicists should remove all theoretical concepts which do not correspond to observable entities. S-matrix theory was Machian while the field theory was not. It is interesting in this context to recall Einstein's letter to his friend the philosopher Solovine, describing how he invented general relativity. Einstein, who at first was a Machian, later argued against Mach's strict view, maintaining that "the mere collection of recorded phenomena never suffices—there must always be added a free invention of the human mind. . . ." He went on to say that the theoretical physicist must be prepared to make an intuitional leap from the experimental data and set up an absolute postulate which itself could not be directly tested but from which one could logically deduce testable consequences. The field postulate, in contrast to the S-matrix theory, represented such an intuitional leap beyond the world of direct experience—a leap which some physicists distrusted.

In retrospect we can now see that the reason field theory failed back in the 1950s and 1960s to give an adequate account of the strong interaction was not that it was wrong but that it was misapplied. The fundamental fields of the strong interactions corresponded not to the hadronic quanta but rather to the quarks and gluons that bound them together. In the mid-1970s, theoretical physicists finally invented a successful field theory of the strong interaction—quantum chromodynamics—based on interacting quarks and gluons. Finding it required not only the experimental confirmation that hadrons were made out of quarks but also developing profound mathematical concepts relating abstract symmetries to the interacting field idea. How did theoretical physicists find the theory of the strong force?

It was not by a direct assault on the problem. Sometimes progress in understanding the problems of physics comes not by directly trying to explain experiments but by exploring new mathematical concepts which are only indirectly related to experiment. Such mathematical explorations are in part undertaken by theoretical physicists, because, as Paul Dirac said, "God used beautiful mathematics in creating the world." Guided by experiment for the details but appealing to the "beautiful mathematics" of symmetry and its relation to field theory for the conceptual overview, theoretical physicists in the 1970s found the field theories of the weak and strong interactions. It is a remarkable fulfillment of the vision that the human mind can know reality and test this knowledge against experience.

In order to understand these recent achievements of theoretical physics we will first have to explore the concept of symmetry. Symmetry has to do with how objects remain unchanged if we transform them. For example, if we take a perfect sphere and rotate it about an axis, it remains unchanged—it has rotational symmetry about any axis. Furthermore, if we rotate a sphere about one axis through some angle and then about a different axis through some other angle, the net result of these two rotations is equivalent to a single rotation about one axis—any two rotations are equivalent to one rotation. Mathematicians can describe these rotational operations in terms of algebraic equations so that the original symmetry of the sphere is now specified algebraically—what is called a Lie algebra, after Sophus Lie, the mathematician.

These ideas of symmetry resulted in one of the most beautiful branches of mathematics, called the Lie group theory. All possible symmetries such as a rotating sphere in various spaces with arbitrary numbers of dimensions were completely classified by the French mathematician Ellie Cartan. Remarkably, these elegant mathematical symmetries have a profound application in field theories of the quantum world.

Chen Ning Yang, a Chinese-American physicist, and Robert Mills, an American, took the first significant step back in 1954, a step which resulted in the revolutionary gauge field theory. They were exploring the geometrical symmetries already developed by the mathematicians—the Lie algebras—and found that if one imposed such a symmetry at each point in space then the existence of a new field was automatically required. This was very remarkable—the imposition of symmetry required a new field, what was called the Yang-Mills or gauge field. "Gauge" means a measuring standard, and the Yang-Mills symmetry meant that one could choose a different gauge at each point in space.

How can a symmetry—like the rotational symmetry of a disk about its axis—require the existence of a field? To understand this without mathematics, imagine an infinite two-dimensional sheet of paper which we are going to paint in various shades of gray ranging from white to black. To decide which shade of gray we are going to use we have a disk which is free to rotate about an axis and is marked around the perimeter of the disk with the numbers 1 to 12, like a clock, so that we can determine how much we have rotated the disk (which has a mark on it like an hour hand). If the disk is set at 12, this means we paint the sheet white; 3 corresponds to a gray shade; 6 is black; 9 is a gray shade again; and returning to 12 we have white again. As we rotate the disk we go through all the shades of gray continuously.

Suppose we set the disk at 4 and paint the sheet a dark gray.

Because the sheet of paper is a uniform shade, if we look at any local region of the sheet we cannot tell where we are on the sheet. Mathematically, we would say that the sheet of paper possesses a global invariance—move around it globally and its appearance does not change. Clearly this invariance is indifferent to the actual shade of gray; we could have as well set the disk at 2 instead of 4.

In the context of this illustration, the idea of a gauge symmetry can be understood rather simply. Suppose we allow ourselves to move over the sheet carrying the disk. As we move we continuously rotate the disk in any way we please, and the position of the disk over each point of the sheet tells us what shade of gray to paint that point. After we finish we look at the result: The sheet is no longer a uniform shade but has all sorts of shades from white to black in different places—it is no longer globally invariant.

However, the lost invariance can be restored if we lay upon the multishaded sheet of paper another sheet of clear plastic which has exactly the complementary shadings that compensate those of the paper—when the paper is darkest the plastic is lightest, and vice versa. The resulting combination is uniform once again—the global symmetry is restored.

In this illustration the sheet of paper corresponds to a quantum field. Recall that we described a quantum field in terms of a 3-D lattice of tiny springs, so we have to imagine shading the springs in three dimensions instead of the two dimensional sheet. Selecting a position of the disk is equivalent to picking a gauge, and rotating the disk as we move from point to point corresponds to what physicists call a "local gauge transformation." Instead of the simple symmetry of a disk one can imagine more complicated symmetries—rotating a sphere about its three different axes selects different colors for the 3-D lattice of springs. The sheet of plastic which restores invariance is the Yang-Mills field—it compensates exactly for the arbitrary freedom of rotating the disk or sphere at each point in space.

Even from our simple illustration we here can grasp the main idea—the Yang-Mills or gauge field is required to restore invariance if we allow ourselves the freedom to perform a gauge rotation at each point in space. For every degree of freedom of the gauge rotation—one for rotations of a disk, three for a sphere—there corresponds a gauge field, so the gauge field may have multiple components. The idea of the Yang-Mills gauge field thus placed the concept of symmetry even prior to that of a field—the gauge fields were a consequence of symmetry.

What did other physicists make of this development? Most theoretical physicists admired the geometrical beauty of the gauge field concept, but they had no idea how it could possibly apply to the world of quantum particles they were trying to understand—to them, it seemed beautiful but useless.

In some ways the Yang-Mills field theory was similar to the nineteenth-century mathematician Bernard Riemann's treatment of geometry which generalized the idea of flat space to curved space. Physicists had no idea how Riemannian geometry applied to physics until Einstein used it in his general relativity theory of gravity.

Physicists wanted nature to make use of the gauge field concept because, like Riemannian geometry, it was based on such a beautiful set of ideas, and nature ought to make use of beautiful mathematics. But there were two great obstacles in the way of applying the gauge theory to physics. The first obstacle was that the gauge field symmetry seemed to imply that the gauge fields had to be long-range and stretch out over macroscopic distances. The only fields that did this were the electric, magnetic, and gravitational fields, and these fields were already explained by known field theories. So where were the gauge fields in nature? The second obstacle was that no one knew how to make a mathematically consistent quantum theory out of the gauge field and apply the renormalization procedure that worked in quantum electrodynamics. New infinities, not covered by the old renormalization procedure, arose when theoretical physicists tried to quantize the gauge field.

It took theoretical physicists twenty years to overcome these two obstacles. But when they were overcome in the early 1970s a new physics revolution began—the gauge field theory revolution. Today physicists believe that all four fundamental interactions—the gravitational, weak, electromagnetic, and strong interactions—are based on gauge fields. The gluons of these interactions are the quanta associated with the gauge fields.

The first problem was why the gauge fields weren't observed if the Yang-Mills symmetry really played a role in physics. Physicists found two ways out of this problem. First they discovered that the symmetry could be broken so that the different components of the gauge field could manifest themselves very differently. This idea of a "spontaneously broken gauge symmetry" was successfully applied to making a gauge field theory of the weak gluons. As a second solution to this problem, physicists discovered that if the gauge symmetry remained exact, then the associated fields remained completely hidden or trapped inside of other quanta. These confined gauge fields were thought to be the "colored" quark-binding gluons of the strong interaction. So the reason that gauge field symmetries are not directly observed in nature was that either they were broken symmetries or, if exact, they would be hidden symmetries.

Progress on the idea of broken symmetries came around the mid-1960s, and the basic idea was rather simple. The equations which described the interaction of gauge fields with other fields had to have the Yang-Mills symmetry, but the solution to the equations did not. Since it is the solutions to the equations that describe the real world, the gauge symmetry did not have to directly manifest itself. Physicists refer to such asymmetrical solutions to symmetrical equations as a "spontaneously broken symmetry"—it is like the symmetry of a human pyramid of acrobats: It is symmetrical but not stable. The natural tendency—what gives excitement to the act—is for the symmetry to break.

It was the idea of a spontaneously broken symmetry that Steven Weinberg and Abdus Salam seized to construct their 1967 gauge field theory of the unified electromagnetic and weak interactions—a theory which became the paradigm for all future unified field theories. Many of the ideas that went into discovering this unified field theory had been previously known, as Weinberg remarked: "Here one is immediately led to an old kind of symmetry which was suggested somewhere in the depths of the past by Schwinger and Glashow, and later by Salam and Ward, and resurrected for this purpose by me in the 1967 paper." For the first time, Weinberg and Salam showed how these ideas of a broken gauge symmetry could be brought together into a realistic field theory with correct experimental consequences. In 1979, Glashow, Weinberg, and Salam shared a Nobel Prize for their work.

The basic idea of this remarkable field theory is that a spontaneously breaking symmetry gives rise to the difference between the weak and the electromagnetic interactions. In the symmetrical situation there are four equally massless gluons. But after a spontaneous breaking of symmetry, only one of these gluons remains massless, and this particle is identified with the photon of the electromagnetic interaction. The other three gluons acquire a huge mass, a hundred times the proton mass. These are the weakly interacting gluons usually called W^+ and W^-, two particles with equal mass that have plus and minus one unit of electric charge, and the Z°, an electrically neutral weak gluon. The differing masses of the four originally massless gluons reflect the broken symmetry. Steven Weinberg summarized this idea:

> Even if a theory postulates a high degree of symmetry, it is not necessary for . . . the states of the particles, to exhibit the symmetry. . . . Nothing in physics seems so hopeful to me as the idea that it is possible for a theory to have a high degree of symmetry which is hidden from us in ordinary life.

How can a symmetry break spontaneously? Abdus Salam gives the following example. Suppose people are invited to a dinner at a round table. Next to everyone's place setting is a salad plate, located between the dinner plates. If you didn't know the dinner rules you could suppose your salad plate is either to the right or to the left of your plate—it is symmetrical. However, if one guest picks the salad plate to his right, then everyone else must follow suit. The right-left symmetry is "spontaneously broken." The symmetry of the Weinberg-Salam model is more complicated, but the idea is similar—the solution to the symmetrical equations is asymmetrical. The asymmetry is responsible for the different masses of the gluons and the difference in strength of the weak and electromagnetic interaction.

The gluons alone could not spontaneously break the symmetry and give themselves different masses. The Weinberg-Salam model introduced yet another quantum called a Higgs particle after Peter Higgs, the theoretical physicist who was one of the first to realize its importance in implementing spontaneous symmetry-breaking. The Higgs particle is like the person who pushes over the symmetrical pyramid of acrobats or is the first to pick a salad plate—its role is to break the perfect symmetry. So besides the weak gluons, W^+, W^-, and Z°, there will have to be the Higgs particles, and most theoretical physicists are convinced that all these hypothetical particles will be discovered once accelerators with the energy capable of creating them are built. That will be part of the experimental physics of the 1980s.

Weinberg and Salam's theory showed physicists how the geometrical ideas of gauge symmetry could be used to solve an important problem of real physics—the unification of the weak and electromagnetic interactions. But when their work was published, hardly anyone paid attention. How could this be? It wasn't ignored because Weinberg and Salam were obscure physicists—they were already well known for other work. The reason their work was ignored was that the second great obstacle to making gauge symmetries work—devising a renormalization procedure—had not been overcome. Many physicists thought that if one started to calculate quantum processes in the model, all sorts of infinites would arise and the theory would be nonsense. This unpleasant situation was soon to change.

The first major breakthrough came from the work of the mathematical physicists Ludwig Fadeev and V. N. Popov in the Soviet Union in 1969. They developed a new and powerful technique for mathematically describing the quantum problem for gauge field theories. Extending their work, a young Dutch physicist named Gerhard 't Hooft in 1971 showed by direct calculations how field theory models of the Weinberg-Salam type were renormalizable—and that was the beginning of the excitement. The formal proof that the Yang-Mills field theory was renormalizable came in 1972 from the Korean-American physicist Benjamin W. Lee, in collaboration with the French physicist Jean Zinn-Justin. These mathematical developments put the renormalization procedure for the Yang-Mills gauge field theories on a par with electromagnetic theory. The final obstacle to constructing realistic gauge field theories had been overcome, and the gauge field theory revolution was underway.

With the gauge field concept firmly in place as a unified theory of the electromagnetic and weak interactions, theoretical physicists now tried to apply it to another of the four interactions—the strong interaction. Experimental physicists had already confirmed that hadrons—the strongly interacting particles—were built up out of quarks. But what bound the quarks together inside of hadrons? Here the idea of gauge fields leaped to the theoretical physicist's mind—why not bind the quarks together with a new set of gluons whose existence was required by a gauge symmetry? The exploitation of this idea led to the creation of the gauge field theory of the strong interactions, called quantum chromodynamics.

The basic idea of quantum chromodynamics is that each quark has a new kind of charge—a "color" charge. Quarks are not really colored—that was just a way of visualizing the three new charges physicists assigned to the quarks. Instead of just a single up quark there was now a red up, a blue up, and a yellow up quark—the three primary colors. Introducing these three extra color charges allowed physicists to postulate a new symmetry among the quarks—a color symmetry. This symmetry was similar to the rotational symmetry of a sphere in a three-dimensional space—each of the three spatial directions now corresponded to one of the three primary colors, red, blue, and yellow. If the sphere was rotated, the different colors mixed, and perfect color symmetry meant that the three primary colors had to be equally mixed. An equal mixture of the three primary colors produces white—not a color at all. Requiring such a color invariance thus implied that only those combinations of colored quarks were allowed that when mixed resulted in no color at all. These colorless combinations of colored quarks (antiquarks are assumed to have the complements of the primary colors) correspond exactly to the observed hadrons. Exact color invariance just reproduced the rules for building hadrons out of quarks!

Now it was clear how to apply the idea of the Yang-Mills gauge symmetry to the strong interaction. The color symmetry of the quarks was postulated to be an exact gauge symmetry, and this implied the existence of eight colored gluons, similar to the photon, that couple to the colored charges of the quarks. But unlike the photon, which because it has no electric charge does not couple to itself, the eight colored gluons *do* interact among themselves. The colored gluons not only stick to quarks, they also stick to each other! The colored gluons are the real origin of the strong interaction.

According to quantum chromodynamics, the color charges of the quarks coupling to the eight colored gluons is the totality of strong interaction physics—all the complexity of the hadrons is to be accounted for by this single idea of a gauge symmetry. The colored gluons provide the binding that traps the quarks inside of the hadrons, so the color charges are to be forever trapped. Quarks, since they are colored, are trapped. The eight gluons because they are colored are also trapped. Only hadrons, which are colorless combinations of the colored quarks and gluons, can exist as free particles, and this is just what we see in the real world. If this idea is correct, and there is ever mounting evidence that it is, then the totality of strong-interaction physics is due to completely hidden forces. The 3-D movie of the hadrons is in black and white; but if you look inside the hadrons at the quarks, it is in color.

We see that nature has indeed made use of the beautiful mathematics of the gauge field symmetry in two different ways. In the gauge field theory of the electromagnetic and the weak interaction, the Weinberg-Salam model, nature broke the exact gauge symmetry in an elegant way. In this case the gauge field quanta —the photon and the weak gluons—can be observed directly. The other way nature used the gauge field theory was in the strong quark binding force, where the color gauge symmetry remains exact but completely hidden. All the colored objects, quarks and gluons, bind themselves permanently together into the colorless hadrons. The invention of these field theories and their application marked the first triumph of the gauge field revolution.

This revolution is an excellent example of the cross-fertilization between pure mathematics and physics. Abstract symmetry principles were invented by mathematicians who could hardly have foreseen their application to the most fundamental physical problems. Other examples of this cross-fertilization can be given— Newton, in order to solve physical problems, invented the calculus, which mathematicians went on to develop. But why is there a relation between mathematics and physics? Mathematics is a human invention inspired by our innate capacity to deal precisely with abstract ideas, while physics is about the material world— something not at all created by us. The connection between our internal logic and the logic of material creation seems gratuitous.

By applying elegant mathematical symmetries, physicists learned a new lesson about the natural world: Symmetries imply interactions. Remarkably, these interactions are precisely the ones observed in the high-energy physics laboratories. The gauge field theory revolution gave physicists a deep clue about the structure of material reality—all the gluons that mediate the interactions are consequences of gauge symmetry. One of those gluons, the photon—light itself—is also a consequence of symmetry. If we could go back to the beginning of time, to the primordial fireball of quarks, leptons, and gluons, when the gauge symmetries were as yet unbroken, instead of *Fiat lux*—"Let there be light"—we might have heard "Let there be symmetry."

11. Proton Decay

. . . We should remember . . . that the basic principles of theoretical physics cannot be accepted a priori, no matter how convincing they may seem, but rather must be justified on the basis of relevant experiments.

—GERALD FEINBERG AND MAURICE GOLDHABER (1959)

MOST PHYSICISTS ENJOY the outdoors. They are mushroom gatherers, bird watchers, hikers, amateur mountaineers whose idea of weekend relaxation is to climb a mountain. Several physics summer schools and research centers are located in or near mountainous areas. I have not seen such uniform recreational impulses in other professions, an observation which prompts speculation on the connection between the nature of inquiry in physics and mountain climbing.

People did not always love the mountains. Just a few hundred years ago the high mountains were regarded as horrible, monstrous places filling people with terror and fear. The inhabitants near them were seen as awful demons, subhumans. But this attitude got transformed into just the opposite, especially by Romantic writers and painters in the nineteenth century. Seen by the Romantics, high mountains became places of impossible beauty, where the quality of light and the expansive solitary grandeur of the high peaks opened the heart of the individual. A man climbing a mountain became the image of self-conscious intelligence pitted against the eternal indifference of the forces of nature. Compared to these forces of nature, we are nothing save for the will that moves our limbs. Only that will is truly our own.

Mountain climbing is an analogue of the research process in

theoretical physics. In working on a problem in physics one is never assured of a solution, because there are many false leads and pitfalls. Likewise in climbing there is no certainty of attaining the summit; the route is often unknown, or sometimes one attains a false summit. But the important point is that if you reach the top, the view is enormous. There is no comparison between what you see from just below the summit and what you can see from the top. Similarly with finally solving an important problem in physics—the view you get is enormous.

In the last decade an important problem has been solved in theoretical physics: the quantization of gauge field theories and the application of these theories to the dynamics of quarks and leptons. This was the gauge theory revolution. A consequence of solving this problem is that physicists have gained a great view of the fundamental nature of matter. From this peak we can see the unification of all the interactions in nature. Whether it is only a false summit time will tell; but right now we can all enjoy the view.

According to these discoveries, the fundamental inhabitants of the universe are the quarks and leptons. The interaction among these is mediated by gluons, which are quanta associated with a field that can be derived from a Yang-Mills gauge symmetry. Four gluons, the photon and the W^+, W^-, and Z°, are responsible for the electromagnetic and the weak interactions among the quarks and leptons. Eight colored gluons are responsible for the strong force that permanently binds the quarks into hadrons.

Theoretical physicists had found two theories to describe the interactions of the quarks, leptons, and gluons: the Weinberg-Salam theory of the unified electromagnetic and weak interactions, and quantum chromodynamics, the theory of the colored quarks and gluons. Both of these theories were based on the gauge symmetry principle. Physicists were therefore tempted to find a single gauge symmetry that incorporated both the Weinberg-Salam theory and quantum chromodynamics—a grand unified theory of the electromagnetic, weak, and strong interactions. It did not take them long to find one. The simplest theory that unifies these interactions was suggested by two Harvard physicists, Howard Georgi and Sheldon Glashow, in 1977, and it is based on a single gauge symmetry of the Yang-Mills type. The crucial idea of their theory, which had already been suggested before their work, was to treat the quarks and leptons on an equal footing before the single symmetry was spontaneously broken. The single Yang-Mills symmetry they postulated gave rise to a set of twenty-four gluons which interacted with all the quarks and leptons in a symmetrical way. This symmetry was then broken in stages. At the first stage, twelve of the twenty-four gluons acquired an enormous mass. They are called "superheavy gluons" and are billions of billions of times heavier than the proton, so heavy that no accelerator will ever create them. The twelve remaining gluons correspond to the familiar four gluons of the Weinberg-Salam model and the eight colored gluons of quantum chromodynamics. The second stage of symmetry breaking follows the pattern of the Weinberg-Salam model, in which three of the four gluons—the weak gluons—get a mass of about one hundred proton masses, while the photon remains massless along with the eight colored gluons. So the end result of this grand unified theory of quarks and leptons corresponds to the world we see.

At first this scheme of unifying the various interactions under the aegis of a single, spontaneously broken symmetry seems like a mere intellectual exercise. The main consequence of the unification of the interactions seems to be the existence of twelve new superheavy gluons that will never be detected, so who cares? But there is an observable consequence of this scheme of unifying the interactions, an important one which has physicists very excited. The superheavy gluons have interactions which destabilize the proton—the main building block of the atomic nucleus.

For a long time physicists thought that the proton was absolutely stable and could not disintegrate into lighter particles. Some even thought that proton stability was a basic principle of theoretical physics—an a priori concept. The reason for the belief in proton stability can be understood from the fact that the proton is the lightest baryon made out of three quarks and has to be stable because there is nothing lighter the quarks can decay into.

The proton is the final remnant of other baryon decays—even the neutron eventually decays into a proton.

But the superheavy gluons changed this, because they did something none of the other gluons did—they could change quarks to leptons. This meant that one of the quarks in the proton could change into a lepton and the proton could now decay. An expected decay mode is for the proton (p) to decay into a neutral pion (π°) and a positron (e^+) according to $p \rightarrow \pi^\circ + e^+$. Because the new superheavy gluons are so incredibly heavy, the probability of such a decay is extremely small—small but not zero. The lifetime of the proton has been calculated by theoretical physicists using the idea of grand unification, and they find that is a factor 1,000 or so larger than the limit of ten billion billion billion (10^{28}) years set by experimentalists who had previously searched for proton decay and did not see it. The prediction of the theoretical physicists that the proton will indeed decay has motivated their experimental colleagues to improve their experiments and look harder for decaying protons. New experiments are now underway that are more exacting, and they will see the proton decay even if its lifetime is a factor of 1,000 larger than the previous experimental limit. If the grand unified gauge theories are right, then these new experiments ought to observe proton decay.

Suppose the unified field theory ideas are right and the experimental physicists will observe proton decay. What does this mean? The most profound implications are for cosmology—proton decay is the death knell of the universe. Most of the visible matter in the universe—the stars, galaxies, and gas clouds—is made out of hydrogen, and the nucleus of the hydrogen atom is a single proton. If protons decay, then the very substance of the universe is slowly rotting away like a cancer that infects matter itself. This rotting away of matter will, according to these unified theories, take a very long time: about a thousand billion billion (10^{21}) times the present age of the universe. We will have lots of time to explore the universe before it vanishes.

The instability of the proton helps to explain a feature of the universe—it is made mostly of matter and not an equal mixture of matter and antimatter. It would appeal to our sense of symmetry if matter and antimatter existed in equal amounts in the universe. But this doesn't seem to be the case. How do we know that far out beyond the visible galaxies there aren't galaxies made out of antimatter? The problem with assuming equal amounts of matter and antimatter in our universe is that although the antimatter galaxies are far away from our galaxy now, according to the big bang theory of the origin of the universe they would have been right on top of us long ago when galaxies first condensed out of the primordial explosion. Matter galaxies would have been annihilated with antimatter galaxies and the result would have been no galaxies at all. So the universe would seem to be made mostly out of matter, not an equal mixture of matter and antimatter.

If the proton is unstable, then it turns out that it is possible to have a matter-antimatter symmetry at the origin of the universe and our sense of symmetry is restored, a point emphasized by the Soviet physicist Andrei Sakharov even before the grand unified gauge theories could explain proton instability. If the proton is unstable, not only does it mean that the proton can decay but also that the reverse process is possible—the proton can be built up out of other quanta. Conceivably, if special conditions are met, more protons might get synthesized in the big bang fireball than antiprotons, thus accounting for the fact that the universe today is mostly made out of protons, not an equal mixture of protons and antiprotons.

The discovery of unified gauge theories of the electromagnetic, weak, and strong interactions has a great impact on cosmology, especially the account of the first few minutes of the creation of the universe. A way to imagine the unification of the three interactions is to go back to the beginning of time and the big bang. The primordial fireball was a mixture of all the quarks, leptons, and gluons at enormous temperatures corresponding to ultrahigh energy. At very high energy the distinction between the weak, electromagnetic, and strong interaction does not exist—they are all unified and have the same strength. The interactions were all symmetrical and equal at the creation, but as the fireball expanded the temperature dropped and the exact symmetry of the interactions became spontaneously broken. With the breaking of the symmetry, the various different interactions became apparent—first the superheavy gluons were distinguished from the ordinary weak gluons and then these were distinguished from the massless photons and colored gluons. Only the massless gluons like the photon are a reminder of that world of perfect symmetry. This symmetry breaking can be thought of as a freezing out of the various interactions as the big bang explosion cooled. Our universe today is the frozen fossil of that remarkable event.

This unification of the interactions, even though it has given us an astounding vista back to the origin of the universe, raises new questions. These are the problems theoretical physicists are struggling with in the early 1980s.

Physicists have no idea why the quarks and leptons exist. By contrast, the gluons could be derived from a symmetry—the Yang-Mills gauge symmetry—which is very fundamental. But no such symmetry exists to account for the quarks and leptons—we don't know why nature has brought them forth. This is Rabi's question all over again—"Who ordered that?" Attempts have been made to unify the quarks and leptons by postulating a very large symmetry, but these symmetry groups are so unwieldy many physicists find them unattractive. As one theorist put it, "If God really picked one of those large symmetry groups, then He is not subtle; He is malicious." The basic problem is that there are just too many quarks and leptons for them all to be fundamental. Perhaps quarks and leptons are not a "rock bottom." The idea that they are composite objects is now being explored, but with no great success.

A related problem is the origin of the quark and lepton masses. The quarks and leptons have different masses, which can be deduced from experiment, but no one has the slightest idea why they have the values they do. For example, the down quark is slightly heavier than the up quark. No one knows why.

We have been describing field theories which unify three interactions—the electromagnetic, weak, and strong interactions. But there are *four* fundamental interactions—what about gravity? Gravity is the mystery man of the four interactions. In spite of the fact that the graviton—the quantum of gravity—is also a consequence of a gauge symmetry, physicists have not succeeded in constructing a realistic unified field theory which includes gravity. Ironically, it was gravity that Einstein wanted to unify with the electromagnetic interaction back in the 1930s. He did not succeed, and this problem is still unsolved. However, there has been recent progress. Physicists have extended Einstein's original theory of gravity—general relativity—to a new theory called supergravity which besides having the graviton has a new gravitational quantum called the gravitino. Currently theoretical physicists are studying supergravity theories to see if they can solve the last (and first) problem of unifying the four interactions—gravity. Possibly new fundamental concepts will be needed before this problem is solved.

Grand unified field theories are still speculative and remain to be tested—especially by proton decay. But if correct, they represent a major accomplishment in the attempt to comprehend the universe. The vista physicists have gained from the mountaintop of these unified field theories lets them see back to the beginnings of the universe and the big bang. Understanding the very origin of the universe is the greatest intellectual challenge of the physical sciences. Grasping that spark from which it all came will take us to the boundaries of physics and conceivably bring us to a new vision of reality.

12. The Quantum and the Cosmos

The effort to understand the universe is one of the very few things that lifts human life a little above the level of farce, and gives it some of the grace of tragedy.
—Steven Weinberg

Origins and destinies are a primary human concern. Every child asks its parents about its birth and death. But even as we attempt to answer such questions we realize that our answers are conditioned by the history of human communities and finally by the natural history of the planet. What is the origin of the earth and sun, and what their end? As we are driven onward and outward in our search for origins and ends, we can ask that question of the stars, the galaxy, and the universe: "Who ordered that? Where did it come from? How will it end?"

Every civilization has addressed these questions, attempting to answer them within the framework of its experience. Answers to these questions are often the subject of myth and the content of religion, as celebrated in the stories of every human community. But ours is a civilization that values the acquisition and exploitation of knowledge as a primary good, and so when we come to ask these questions we turn to science. Here we learn that what experimental science can teach us about the universe is limited by available technology. We can know only what our instruments can reveal.

Since the Second World War there have been at least two major technological developments that bear on these questions. First is the deployment of radio telescopes and their auxiliary electronic systems that explore components of the electromagnetic radio spectrum previously invisible. Second is the advent of nuclear physics and elementary-particle physics as an experimental science. These advances in technology and science have done much to bring us to the current view of the origin of the universe, so much so that ten years ago it was impossible to tell the story of the big bang the way we presently understand it. It is safe to say that we have learned more about the origin of the universe in the last decade than in the previous centuries.

The way scientists approach the problem of the origin of the universe can be appreciated if we imagine a courtroom trial. The reason a trial is needed in this branch of science is that there is only one universe and its creation is unique. Scientists cannot go out and test their theory on the spot, because the event is over. The origin of the universe is like a crime that happened in the past—it is no longer happening, and all scientists can do is assemble evidence pointing to the event and draw the best conclusions possible. The judge might be some elder statesman of science who has no stake in the outcome of the trial. The jury consists of representatives of the various scientific professions. What is on trial is various views and theories about the origin of the universe. Lawyers, mostly theoretical physicists and astrophysicists, present their cases defending a specific view of the creation, calling on witnesses, experimentalists who present data.

Some people claim that there is no need for a trial at all—there was no creation of the universe and it always existed pretty much in the state we see it today. This viewpoint, once widely held, expresses the steady-state model of the universe—there is no beginning and no end; the universe is in eternal equilibrium. The reason that the steady-state model could be maintained a decade ago was that there really was very little evidence about the origin of the universe. But today that situation has changed dramatically.

The present view of the creation, what became called the "standard big bang model," maintains that the entire universe originated in an enormous explosion. All matter, the stars and galaxies, was once concentrated into a very confined region in a primordial matter soup. This matter soup expanded rapidly—it exploded. In so doing it cooled down, enabling nuclei, then atoms, and finally much later galaxies, stars, and planets to condense out of it. This explosion is still going on today, except that the universe is much colder now as it expands. Contrary to our impression of the changeless heavens, the universe was and continues to be a place of great change.

There are two pieces of experimental evidence upon which the big bang cosmology rests its case.

The first is the discovery of the expansion of the universe by Edwin Hubble in 1929–1931. He observed that the red shift of the light from distant galaxies is proportional to their distance from us. His conclusion is based on the fact that an atom which is moving away from us at high velocity, such as in a distant galaxy, has its spectral lines shifted to the red in proportion to its velocity —a Doppler shift, like the shift in the frequency of the sound of a train's horn as it moves away. Since the red shift is proportional to the velocity, it follows that the velocity of a distant galaxy and its distance from us are also proportional. The uniform expansion of the universe is certainly the simplest conclusion one can draw from Hubble's data. All other interpretations require a new exotic effect for which there is presently no evidence.

The second major experimental finding is the microwave background radiation discovered by Arno A. Pezias and Robert W. Wilson in 1964. These researchers at Bell Laboratories found that the black empty space of the universe was not absolutely cold; it has a slight temperature of three degrees Kelvin* above absolute zero. This temperature is due to a radiation bath of photons that permeates all of space. The distribution of frequencies or colors of those photons has been measured, and it is just that of Planck's black-body radiation curve for a black body at a temperature of three degrees Kelvin. In this case the black body is the whole universe.

---

* The Kelvin scale of temperature, unlike the familiar Fahrenheit scale, begins at absolute zero—the lowest temperature—so that 0 degrees Kelvin is about *minus* 460 degrees Fahrenheit.

The interpretation of this radiation bath is that it is heat left over from the big bang—it is like observing the heat of rocks around a campfire and concluding that there was a fire burning there not long before. Once the universe was a highly concentrated matter soup at temperatures of billions of degrees. Then it exploded, until today it has cooled down to a few degrees as a consequence of the expansion. Its temperature is still dropping—but very slowly now. The evidence of this microwave background radiation convinced most scientists sitting in the jury of the correctness of the big bang model. The universe was created from an explosion—it did not exist for all time.

Astrophysicists and cosmologists have constructed a theoretical model of the creation of the universe. They start their clocks at about the first hundredth of a second after the creation, because before the first hundredth of a second, the temperatures were so high and the energies so great that they must extrapolate beyond today's high-energy physics theory—it is very speculative. After the first hundredth of a second, physicists think they understand the physics that describes the expansion sufficiently well that they can say with some certainty what the situation was.

At the first hundredth of a second the temperature of the primordial soup was one hundred billion degrees Kelvin, a hot soup indeed. The soup consisted mostly of electrons, positrons, photons, neutrinos, and antineutrinos. These particles were continually being created and destroyed as they interacted. The density and temperature of this soup was so great that it was as likely that an electron and positron annihilated into photons as that photons collided to create an electron-positron pair. In addition to these electrons, neutrinos, and photons, there was also a small contamination of protons and neutrons, about one billionth the number of photons, in the primordial soup. Hang onto that small speck in the soup, because out of that speck, all the galaxies and stars and ultimately the earth will be made.

After the first tenth of a second had passed, the universe cooled down to about ten billion degrees Kelvin; after fourteen seconds, down to about one billion degrees Kelvin. This was cool enough to take the electrons and positrons out of equilibrium with photons and neutrinos, and now if positrons got annihilated they wouldn't get recreated—all that remained was electrons, neutrinos, and photons. After three minutes had passed, the temperature of the universe dropped sufficiently—the particles were less agitated—for the small contamination of protons and neutrons to combine into nuclei. The first nuclei formed were the lightest ones, deuterium and helium. Using the laws of nuclear physics, physicists can calculate the amount of helium and other light elements made in this way, and they find that the amount of helium made in the big bang is about 27 percent of all the matter in the universe, in good agreement with observational evidence. These calculations and the agreement with observations lend great credibility to the big bang model.

Only after about a hundred thousand years had elapsed—the universe was getting quite old—did the temperature drop sufficiently for the electrons to combine with the nuclei to form atoms. Great clouds of atomic matter emerging from the explosion began to condense into galaxies and stars. Inside the stars, the heavier elements like carbon and iron got cooked up from hydrogen and helium by a process called nucleosynthesis. After a few billion years, the universe began to look as it presently does. Today it is between ten and twenty billion years old. The earth, by contrast, is about four to five billion years old, and life on earth about two and a half billion years old.

Everything you see about you is a fossil. Just as deep rocks are fossils of the creation of our planet, so the nuclei and atoms are the fossils of the big bang. There was a time that they were created; they did not exist for all time. We are a fossil world, frozen at very low temperatures relative to the temperature of the primordial matter soup that gave birth to everything.

This view of the universe has some serious critics, but they mostly criticize the details, not the idea itself. Like Copernicus's heliocentric model of the solar system given centuries ago, the big bang model seems to be substantially correct. With new experimental technology about to be available—the very large radio antenna array in New Mexico and the space telescope—the "standard big bang model" can be further tested. We may be in for some surprises, but it would be very remarkable if the major features of this creation story changed.

In spite of the very satisfying qualitative and quantitative account that the big bang model gives of the present universe, physicists are tempted to look beyond the first hundredth of a second. Here they must speculate on the basis of what they know from high-energy physics and the ideas of unified gauge field theories.

It seems certain there were protons and neutrons in the primordial soup at the first hundredth of a second, a tiny contamination. At the first millionth of a second they did not exist either. The protons and neutrons are themselves fossils, frozen-out mixtures of colored quarks and gluons. In the first millionth of a second, physicists speculate, the soup consisted of the fundamental particles that we know today, the leptons, the quarks, and the gluons, all interacting with each other. At still-higher temperatures and earlier times the quarks and leptons may have transmuted into one another—interactions which in those hot times would have created the matter-antimatter asymmetry of the universe and today be manifested as proton instability. At the highest temperatures and earliest times all distinction between the interactions is lost—it is a universe of perfect symmetry.

How did this big bang happen? What was the origin of the primordial soup of quarks, leptons, and gluons? Where did *that* come from? Certainly this is not a question physicists can confidently answer on the basis of experiment and theory. However, we can speculate. There is an answer which I am attracted to based on the rules of physics as we now know them. The answer to the question "Where did the universe come from?" is that it came out of the vacuum. The entire universe is a reexpression of sheer nothingness. How can the universe be equivalent to nothing? Look at all those stars and galaxies! But if we examine this possibility carefully we learn that the universe, even in its present form, could be equivalent to nothing.

A remarkable feature of the present-day universe is that if you add up all the energy in the universe it almost adds up to zero.

First there is the potential energy of the gravitational attraction of the various galaxies for each other. This is proportional to the mass of the galaxies. Since one must supply energy to push the galaxies apart, this counts as a huge negative energy in our energy bookkeeping. On the positive side of the ledger is the mass energy of all the particles in the universe. This adds up to another huge number, to about a factor of ten smaller than the negative energy. If the two numbers matched, the total energy of the universe would be zero and it wouldn't take any energy to create the universe.

Astronomers are searching for the "missing mass" that would make the total energy zero. There are lots of places this missing mass could be hiding. Galaxies have most of their mass in large halos that are invisible. Perhaps there are large, invisible black holes at the cores of galaxies. The most recent candidate for the missing mass is a small neutrino mass. The universe could be filled with massive neutrinos, and that could be the major part of the mass. It is hard to tell now, but it is conceivable that some large mass-energy has been overlooked, and then the total energy of the universe could be zero.

Until recently, an obstacle to the idea that the universe was created out of a vacuum was explaining where the protons in the universe came from. But with the theoretical possibility that the proton is unstable, this objection is overcome. The present matter-antimatter asymmetry of the universe does not reflect the original state of the primal fireball, which could have perfect symmetry. Therefore it seems possible that all objections to the idea that the universe is a representation of the vacuum can be overruled. But how can the vacuum spontaneously turn itself into a fireball of quarks, leptons, and gluons—the big bang?

It seems that a vacuum is stable. Likewise it once seemed that atoms were stable, but now we know that they are not—the nuclei of atoms can disintegrate spontaneously in a spectacular reaction which is manifested as radioactivity. According to the laws of quantum theory there is a probability that otherwise stable nuclei will decay. I believe it is possible that a vacuum is similarly unstable—there is a tiny quantum probability that a vacuum will convert itself into a big bang explosion. No explanation exists for a specific nuclear decay—only a probability can be given. Similarly, no explanation would be needed to account for the specific event of the big bang if this idea is right. Since no one is waiting for the event to happen, even if it has an infinitesimal but finite probability, it is certain to happen sometime. Our universe is a creation by the God that plays dice.

Although scientists argue over the details of the beginning of the universe, a consensus about its general features has been reached. If we now turn to examining the end of the universe we find no such consensus exists: Were we to go back to our courtroom of scientists, we would find no agreement. While the birth of the universe left scattered clues as to what happened long ago, no firm clues exist yet for what is about to happen. Discussing the end of the universe is like speculating on a crime that is about to happen. You might even find some clues that a crime was about to happen, but no number of clues implies that the crime must happen. The most that scientists can do is to gather evidence of the event that is going to happen and then assemble scenarios that are consistent with that data.

Physicists find there are basically two ends to the universe—fire or ice; we will either be fried or frozen. These two scenarios harken back to Alexander Friedmann's solutions to Einstein's equations in which he showed that the universe is either closed or open. The open universe is one in which the universe, now expanding, continues to expand indefinitely. The closed universe expands out to a definite limit and then recontracts.

Whether we are living in an open or closed universe is an experimental question that can be answered once the data exist. The observed red shifts of distant galaxies follow Hubble's law—the red shift is proportional to distance—out to the farthest galaxies. This means the universe is expanding at a uniform rate—it is not slowing down or speeding up. If the expansion is speeding up the universe is open; if it is slowing down the universe can be either open or closed.

There is another way to get at the answer to the question of whether the universe is open or closed, and that is by accurately determining the total mass density of the universe. Right now there does not seem to be enough mass to slow the expansion, and we would conclude that we are in an open universe. But there may be "missing mass"—invisible matter—that could change the conclusion.

In the closed-universe scenario, the universe will continue to expand for perhaps tens of billions of years. Then the expansion stops and the contraction begins. The distant galaxies instead of being red-shifted in their light will now become blue-shifted. After billions of years the sky will grow hotter and hotter. The big bang 3-D movie is now run in reverse, and eventually everything collapses into the matter soup it was at the creation. Whether the universe "bounces" at this point and reexpands depends on physics which is not well understood. But it is unlikely that humanity or whatever it becomes can survive the collapse or bounce. If the universe is closed, our end will be destruction by fire. Some people think that economics is "the dismal science," but I think it is cosmology.

In the open-universe scenario, the universe continues to expand indefinitely and the galaxies move farther apart. At first this seems to be a mild alternative to death by fire. But it is clear that the universe, even if it is open, cannot remain as we know it today. There are physical processes, either known or conjectured, that imply the deterioration of the universe if it is given the time to age gradually. We have already alluded to the fact that experiments are currently underway to examine if the proton is unstable. If proton decay is observed then it means the end of the universe as we now know it, in about the lifetime of the proton. Attacked by a cosmic cancer, the whole universe will rot away.

Even if the proton is much more stable than our current theories suggest, there are other disasters that can take place. The universe is indeed a treacherous place. Low-mass stars cool off in about one hundred thousand billion (10^{14}) years and planets detach from stars, by collisions with other stars, in about a million billion (10^{15}) years. Galaxies have an upper limit to their lifetime, and their high-velocity stars will fly off in about ten billion billion (10^{19}) years. The remainder presumably get swallowed by large black holes in the galactic core.

Black holes may play an important role in the end of the universe, because most of the matter we see today may end up in black holes. But the modern theory of black holes devised by the English physicist Steven Hawking implies that even black holes are unstable and radiate energy. In one scenario, the end of the universe consists of distantly separated black holes and long-wavelength electromagnetic and gravitational waves, forms of energy from which nothing of interest can be made. This would be the ultimate energy crisis—a cold cruel world which ends "not with a bang but a whimper"—a big whimper.

With either scenario, fire or ice, the human species, if it does not first eradicate itself, will have a long time to think things out. Only a decade ago physicists and astrophysicists had to apologize for thinking about the beginning and end of the universe, because in the absence of hard experimental data it was necessarily speculative thought. Today that situation has changed; there are data and we will know even more in the near future. The discovery of the laws of quarks, leptons, and gluons and the advances in astronomical instruments provide powerful tools to unravel the puzzle of the universe.

It is futile to develop either an optimistic or a pessimistic attitude toward the problem of the end of the universe. I realize that it is difficult to keep from projecting our desires onto the universe; even the most intelligent people do it. But optimism, the belief in our capacity to survive, is programmed in us by an evolutionary process only a billion years old, and conditioned by earth's environment. It may not be appropriate for the eons of time that face the species. Those endless reaches of time will condition life in unknown ways.

Physicists do not yet know if there really exist ultimate laws that express the final conditions of all existence. Perhaps there is no absolute law which governs the universe and life in that universe.

Until the final chapter of physics is written we may be in for lots of surprises. Conceivably, life might be able to change those laws of physics that today seem to imply its extinction along with that of the universe. If that is so, then might not life have a more important role in cosmology than is currently envisioned? That is a problem worth thinking about.

In fact, it may be the only problem worth thinking about.

PART III

The Cosmic Code

Heaven wheels above you
Displaying to you her eternal glories
And still your eyes are on the ground.
—DANTE

1. Laying Down the Law

It is not your part to finish the task;
yet neither are you free to desist from it.
—Rabbi Tarfon, *Pirke Avoth*

Many years ago, camping near timberline in the High Sierras of California, I watched the stars appear above the mountains. As I fell asleep my eye caught a patch of haze in the night sky; I soon realized this was the galaxy Andromeda, the beautiful spiral of swimming suns. It was a fuzzy patch, the only galaxy out of millions that can be seen by the unaided eye—I was looking across intergalactic space. Hours later I awakened to find that Andromeda along with the familiar constellations had shifted in the night sky. Night after night I saw a periodic movement of the stars.

The celestial motions, in contrast to the capricious turns of human life and the play of social fashions, proceed with serene certainty. It is no accident that ancient priests searching for order on earth looked to the stars. They realized that the profound message of the definite motion of the stars is that certain knowledge of the universe is indeed possible. By tracking the planets, the sun, and the moon year by year they learned that their motion was not random but has a pattern—there is order in heaven. And this celestial order can be used to determine the seasons and the annual flood of rivers like the Nile—observations which gave birth to the idea of physical law. Beyond the world of capricious appearance lay another world which could be ordered by our minds.

Looking for the natural laws is a creative game physicists play with nature. The obstacles in the game are the limitations of experimental technique and our ignorance, and the goal is finding the physical laws, the internal logic that governs the entire universe. As scientists search for natural laws the ancient excitement of the hunt fills their minds; they are after big game—the very soul of the universe.

What are the physical laws? How do we know what we are looking for? The ultimate answer to these questions cannot be given—it is still the goal of the physicist's search. But the fundamental laws that govern most of the world of ordinary experience are known. If we compare the exploration of reality to the exploration of the earth then we could say that physicists have already explored the lush green valleys and meadowlands. Today they are exploring the deserts of reality, areas far removed from immediate human experience—the beginning and end of the universe and the world of subatomic particles. They do not know if they are near their ultimate goal of finding the ultimate laws of nature—that depends on what they discover.

Although physicists do not yet know the ultimate laws of nature or even what form they will take, over the centuries they have found characteristics—almost definitions—of physical laws. These are of interest because they give us clues for what we are looking for—not in detail but in broad outline. These characteristics of physical laws are not arbitrary but reflect the relation of our mind to the world it attempts to grasp. I will describe a few of these features of physical laws such as their:

1. Invariant nature
2. Universality and simplicity
3. Completeness
4. Relation to observation and experiment
5. Relation to mathematics

Let us look at each of these characteristics in turn.

1. THEIR INVARIANT NATURE

A physical law is a proposition stating that something always remains the same—an invariance. Action is always equal to reaction; the speed of light in empty space is always an unchanging constant; total energy is always conserved. Physical laws are therefore unlike social "laws" which simply stipulate invariances. The difference between social law and physical law is the difference between "thou shalt not" and "thou cannot." No one will go to jail for violating the law of energy conservation. But it is not obvious that there are any physical laws, true invariances, of the natural world. In the world we see nothing but change, often chaotic change. Why then is it reasonable to suppose that all this change is subject to law?

Reflecting on this problem, Newton had the crucial insight of conceptually separating the actual state of the world—which can be very complicated—from the laws which describe how such a state changes—which can be quite simple. The complicated appearance of the world is inessential to understanding the invariances—the physical laws—which give a detailed description of how the world can change.

The idea that beyond the changing world is a changeless one is very remarkable. Think of a disc that is rotating about its axis. As it moves its appearance remains the same, because a disc has a symmetry—rotations about its axis leave it unchanged. This illustrates the modern idea of invariance or physical law—it is a consequence of symmetry. If an object has symmetry like the disc it means we can move it without changing it—symmetry implies invariance. That is why physicists are always searching for symmetries. They know if they find a symmetry it implies a new invariance—something that cannot change.

The easiest way to understand symmetry is in terms of a symmetrical object. But the notion of symmetry can also apply to ordinary space. If I take anything and move it across empty space I cannot tell the difference. This translation of an object is like rotating the disc—a transformation that leaves the physical state unchanged. Translational invariance is a symmetry of ordinary space, and it means that the laws of physics apply to events independent of their location in space. Anyone who has traveled long distances on the interstate highway system knows about translational invariance. One can spend a night in a motel, travel five hundred miles, and spend the next night in an identical motel room.

Another symmetry or invariance of the laws of physics is time translational invariance. The results of measurements of quantities like the charge on an electron or the strength of gravity should not depend on whether the measurement is done on Monday or Wednesday. But what is the significance of these symmetries of space and time? They seem simple and yet they have a profound consequence.

The mathematician Emmy Noether discovered the full consequences of symmetry in physics only in this century. She showed that for every symmetry of physics—like the translational symmetries of space and time—there is a conservation law. An example is the conservation of energy. If we measure the total energy of a closed physical system—the sum of all energy of motion, potential energy, heat energy, chemical energy, and so on—the total energy remains unchanged, although one form of energy, like energy of motion, may be converted to another, like heat energy. The conservation of total energy is a precisely verified fact. What is remarkable (and this was the point of Noether's work) is that the conservation of energy is a logical consequence of the time translational symmetry.

How is that possible? What does the invariance of physical laws from Monday to Wednesday have to do with energy conservation? In order to understand the answer to these questions I will assume, to the contrary, that the laws of physics can change in time; in particular I will assume that the law of gravity can change. By assuming the impossible we will understand the possible. Using this changing law of gravity I will then show how to build a perpetual mobile, a device that generates energy for free and thus violates energy conservation. The conclusion is that since this is impossible, the premise I began with—that laws of physics are not time translation invariant—must be false. The unchanging nature of physical laws in time logically requires the conservation of energy.

Imagine a water wheel which conveys water from a high reser-

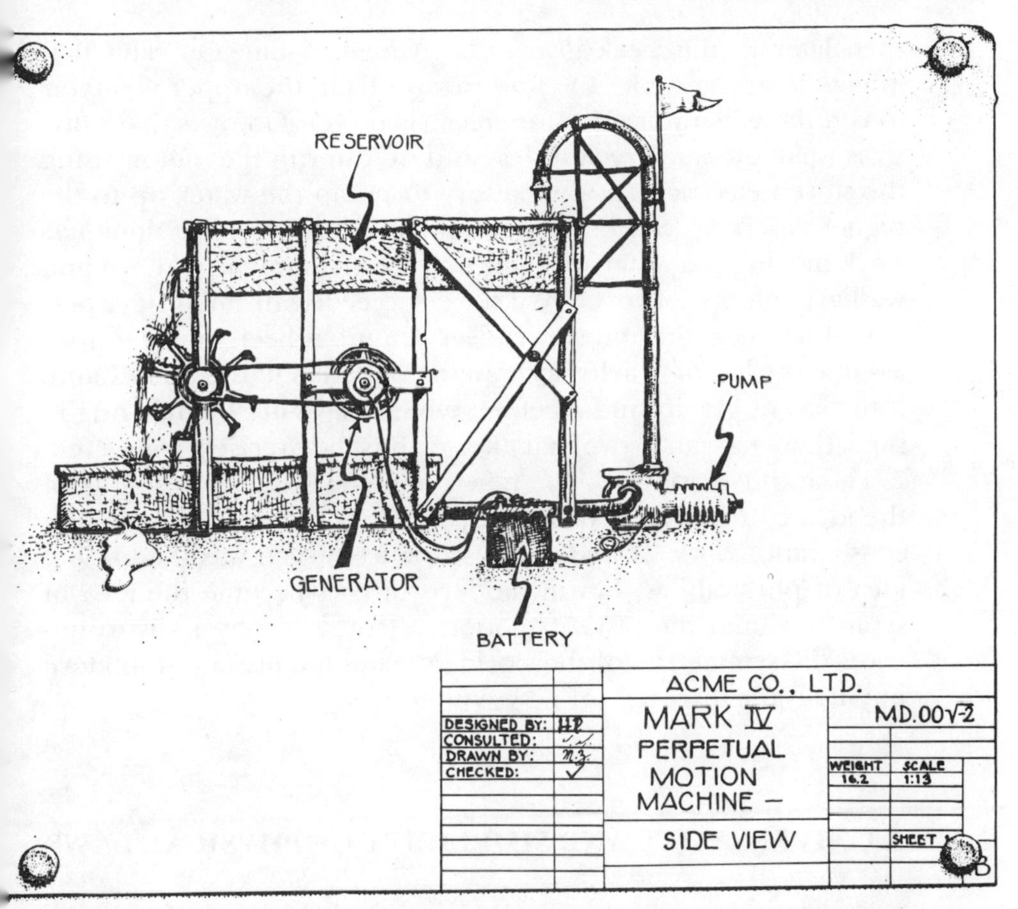

A design for a perpetual motion machine and a solution to the world's energy needs. Such a machine would actually work if the laws of physics were not time translation invariant—if the basic laws could change from day to day. This illustrates Noether's theorem: An invariance of the laws of physics such as time translation implies a conservation law such as energy conservation.

voir to a low one and by running it backward from the low one up to the high one. A water wheel is attached to both an electric motor and an electric generator which is hooked up to a battery that stores electricity. We assume that the law of gravity changes in time—on Monday gravity is stronger than on Wednesday, and then later in the week gravity is stronger again. On days that gravity is strong we let the water down from the upper reservoir to run the generator and store electric power. On days that gravity is weak the water weighs less and we can run the motor, using the stored electricity in the battery to pump the water up to the higher reservoir. Since the water weighs less, we have done less work moving the water up than we received moving it down and we have energy left over. We have succeeded in building a perpetual mobile—thus energy conservation has been violated if we assume the law of gravity is changing in time. On the other hand, if the law of gravity and all other physical laws are unchanging in time, then one can prove that energy must be precisely conserved.

Today physicists look for new symmetries, generalizations of the idea of space-time symmetry, knowing that these imply new conservation laws. The idea of invariance, which was the ancient idea of physical law, has in modern physics become the idea of symmetry, and the task of modern theoretical physics is to uncover the symmetries of the world. Most of the history of modern physics is the discovery of new symmetries.

2. UNIVERSALITY AND SIMPLICITY OF PHYSICAL LAWS

Many years ago I asked T. D. Lee, a Nobel laureate in physics born in China, about his educational experiences before he went to Chicago to study with the physicist Enrico Fermi. What had impressed him as a student in China when he first encountered physics? Without hesitation Lee replied that it was the concept of universality of physical laws that had struck him most deeply—the idea that physical laws applied to specific phenomena here on earth, in one's living room as well as on Mars, was new and compelling to him.

An example is Newton's law of universal gravitation, a law that is not just true at certain moments, but is indifferent to time. Furthermore, Newton's law unified gravity on earth with gravity in heaven; it's indifferent to place. His deep insight was grasping that the same rule applied to the orbit of the moon about the earth as to an apple falling in his mother's garden. I like to imagine that Newton, sitting in his mother's garden, also saw the morning moon and realized that it too was falling, like the apple, toward the earth. Only the centripetal force due to the motion of the moon in orbit keeps it from crashing into the earth—stop the moon and it would fall like the apple. It took thousands of years to prepare a mind for this insight. We know today that Newton's law also holds with great accuracy for the motion of galaxies, light-years in diameter—the law of gravitation is universal.

The universality of physical laws is perhaps their deepest feature—all events, not just some, are subject to the same universal grammar of material creation. This fact is rather surprising, for nothing is less evident in the variety of nature than the existence of universal laws. Only with the development of the experimental method and its interpretative system of thought could the remarkable idea that the variety of nature was a consequence of universal laws be in fact verified.

The word "theory" comes from the Greek "to see." The activity of the theoretical physicist is to perceive the internal logic of nature. His interpretations of nature are called theories; they are pictures of the material world made to render it comprehensible, and to succeed in this they must be simple.

The idea of the simplicity of the laws of physics is not easy for an outsider to grasp, because physics seems so complicated. But the remarkable feature of physics is that all complications arise in a logical fashion from a few elemental but profound concepts, as a tree from a single seed. It may take years of study for a student to grasp the simplicity of the core concept of the fundamental laws. Even for the research physicist the realization that such simplicity emerges after the struggle is over is part of his deepest conviction. As Einstein remarked, "The aim of science is, first, the conceptual comprehension and connection, as *complete* as possible, of the sense experiences in their full diversity, and, second, the accomplishment of this aim *by the use of a minimum of primary con-*

cepts and relations (seeking the greatest possible logical unity in the world picture, i.e. logical simplicity of its foundation)."

3. COMPLETENESS

In the Cabala, a collection of Jewish mystical writings, the *yetzer harah,* the "evil impulse," is the sin of desiring completeness. Only God knows completeness; the human attempt to imitate God and achieve completeness in any endeavor is a sin. Yet physicists strive for completeness, because they know a great theory cannot be a partial picture of nature but must give the complete laws of a class of events. For example, general relativity theory, the modern theory of gravity, describes all gravitational effects, not just those for weak gravity fields. Physicists' ultimate goal is to have a unified theory of all physics.

Historically, the various branches of physics, each dealing with a definite aspect of nature, have grown together. Electricity and magnetism, once seen as separate physical forces, became unified in Maxwell's electromagnetic theory. Space and time are unified in Einstein's relativity theory. Today theoretical physicists working in quantum field theory have found "grand unified field theories" unifying the strong nuclear force and the electromagnetic and weak forces, and are currently striving to incorporate the force of gravity into this unification. Should they accomplish this goal it would be the completion of physics as we know it now. This ancient dream of reason seems almost within our grasp.

While today it is conceivable that physics will realize its dream and come to an end, I doubt it. There is little chance that physicists will have a complete theory of *all* nature in the near future, although most things in our immediate experience can be accounted for. The result of every unification in physical theory is to climb up to a new step and from this height see a new vision of nature. The provincialism of our previous position becomes apparent and new questions can be asked—and as long as profound questions can be asked we know we have not completed our work.

4. RELATION TO OBSERVATION AND EXPERIMENT

Since the time of Francis Bacon in the early seventeenth century, "science" has meant experimental science. Bacon popularized the idea that the only way to study nature scientifically is to do experiments. An experiment is a controlled experience in which the conditions of the experience are systematically altered. It is one step beyond simple observation, which is merely passive and in which no attempt is made to alter the conditions of the experience. Passive precise observation is the first step, active experiment the second.

Before the importance of precise observation was appreciated, medieval European physiologists, interested in the classification of mammals, knew of the existence of elephants. But they had never seen one. The sheer size of these large creatures raised the question of how elephants copulated. Imaginative writers suggested various solutions. Perhaps elephants backed into each other, or perhaps they went underwater where the enormous weight would be relieved. One writer suggested that the bull elephant dug a large hole for the female. Strabo, the geographer, writes that the male elephant, copulating in a frenzy, impregnates the female by "discharging a kind of fatty matter through the breathing tube he has beside his temples." None of these writers had the opportunity to observe the beasts, who manage the act with considerable grace. The exercise of the imagination is its own reward; but it is not scientific observation.

Physical theory without experiment is empty. Experiment without theory is blind. It is the experimentalists who keep the theorists honest. They realize that ruthless honesty, dogged persistence, patience, openness, precision, and luck may pay off in the discovery of new natural phenomena. Experimental physicists discovered radioactivity (the decay of the atomic nucleus into other particles), the photoelectric effect (the emission of electrons when light hits a metal plate), atomic line spectra (the distinct colors of light emitted by radiating atoms), and particle-scattering experiments which pushed the theorists into the invention of the quantum theory. These discoveries could not be explained by the older Newtonian physics, and between 1900 and 1926 it became apparent that a new physical theory was needed. The invention of the new quantum theory shows that transformations of thought do not come of their own accord but are impelled by empirical circumstance. I am impressed by how often our minds resist new ideas, yet when willing or compelled are capable of rapid adaptation.

What is the relation between theory and experiment? It is not merely that the theorist proposes a hypothesis and the experimentalist verifies or destroys it, though it can be that. But more often the experimentalist discovers a whole new reality. Examples are the discoveries that opened up the atomic world, radioactivity, the photoelectric effect, and atomic line spectra. The theoretician must then make a leap of the imagination, combining new data with new theoretical ideas. The new theory may in turn suggest experiments that crucially test its hypothesis. The relation between theory and experiment is like a dance in which sometimes one partner leads and sometimes the other.

In astrophysics, the theoreticians are by necessity ahead of the experimentalists. They make theoretical models of the interiors of stars, the cores of galaxies, star systems with black holes, and the first seconds of the big bang. These parts of the universe are not easily accessible with current technology, and observational evidence is hard to come by. With the future deployment of very large radio antenna arrays and the space telescope, new information about the structure of such astrophysical objects will be at hand. Meanwhile the astrophysical theorists are hungry for experimental data.

In the case of high-energy physics, by contrast the experimentalists are often ahead of the theorists. There are enormous amounts of experimental data on the scattering of protons and electrons producing exotic new forms of matter—data which is only partially understood. We theorists believe we have a fundamental theory that can explain all or most of it. Here the problem is not that the theory or the data do not exist but that the mathematical complexities of the equations are so great that no one has been able to make contact between the theory and experiment except at a few places, and there theory and experiment agree. Perhaps the very powerful electronic computers of the 1980s will enable physicists to make a detailed comparison of the recent theory of elementary particles and experiment.

The relation between theory and experiment is symbiotic. Theory provides the conceptual framework that renders experiment intelligible. Experiment moves the theorists into a new realm of nature that sometimes requires a revision of the very concept of nature itself.

5. RELATION TO MATHEMATICS

I have always been fascinated by ambiguity. In our emotional life we use ambiguity for the purpose of avoiding personal confrontations, at least temporarily. In emotional life ambiguity may simply be undesirable, but in science it is a disaster! This is one reason why physical laws are expressed in the precise language of quantitative mathematics. Mathematics makes the theorist's statements unambiguous and hence they can be disproved by an experiment.

An important feature of modern theory is not that the conclusions are provable but that they must be disprovable. A theory may be founded on very general laws. But from those laws we must be able to deduce extremely specific properties of the world, like the motion of an electron in a magnetic field. Only specific, unambiguous predictions can be tested. A theory cannot be right in general unless it can also be wrong specifically. This epistemic vulnerability lies at the heart of an experimental science; indeed, the essence of the scientific method is the willingness to place one's own ideas in jeopardy.

To be falsifiable a theory must be logically precise and unambiguous. Otherwise it cannot even be wrong! In order to possess that unambiguity, physics is specified in the precise language of mathematics. As Heisenberg remarked, "Science has given an important example of the fact that an extraordinary extension of the most abstract fundamentals of our thought is possible without our having to accept any lack of clarity or precision."

Why is it even possible to formulate the laws of nature in terms of mathematical equations? Eugene Wigner, a theoretical physicist at Princeton University, in an essay called "The Unreasonable Effectiveness of Mathematics in the Natural Sciences," dwelled on the peculiar relation between mathematics and physics. The physical world is quantifiable, which means that we can measure it and assign numbers to it. An example is Boyle's law, discovered three centuries ago. We can measure the enclosed volume V, the pressure P, and the temperature T of a gas. But why should there exist an algebraic relation such as Boyle's law, $P \times V = T$, which states that the product of the pressure times the volume is equal to temperature? Such formulas demonstrate the remarkable success of mathematics in describing physical phenomena. But the basis for this success is not clear. "One cannot escape the feeling," wrote Heinrich Hertz, a nineteenth-century German physicist, "that these mathematical formulae have an independent existence and an intelligence of their own, that they are wiser than we, wiser even than their discoverers, that we get more out of them than was originally put into them." Formulas expressing laws of nature impose constraints on the natural world, maybe even on God. Einstein said one reason he worked on physics was to determine whether or not God had any choice in creating the universe the way He did.

Physicists used to believe that they could capture all of nature in their net of mathematics. Everything that happened, down to the finest detail, could be determined. But, remarkably, in the modern quantum theory the idea of a mathematical description of all of nature has broken down. Individual quantum events, such as the radioactive disintegration of one nucleus, are not subject to any mathematical-physical law; only the distribution of these events, averages over many occurrences, are subject to the laws of quantum theory. The laws of physics are not deterministic but statistical, a discovery which implies the end of a mathematical description of all of nature.

Physics brings a wonderful comprehensibility to the cosmos. How can we comprehend a universe that we did not invent? The philosopher Kant thought that the internal logic of nature corresponded to the internal logic of the human mind, and this was why nature was comprehensible. If there is a correspondence between nature and the mind, it is presumably not an accident of biological evolution. The human brain, for all its magnificence, is nonetheless the product of evolution; it is part of nature and subject to physical laws.

What are the limits to the idea of comprehending the laws of nature? We can imagine building an artificial intelligence, a big computer, whose job it is to discover the laws of nature. But the operations of the computer are themselves limited by the laws of nature. For example, because the speed of light is finite it takes a finite time to transmit information from one side of the computer to the other. This limitation on any real computer may limit its operations in a fundamental way and prevent it from finding the physical laws of nature that also determine its own operations. Does this fact constitute a limitation on its logical operation? If so, it may provide a boundary to our knowledge of the laws of nature. I think that the problem of determining the limits to knowledge as a consequence of the material structure of any real intelligence will be part of the physics of the future.

In reflecting on the relationship between physical laws, theories, and the natural world, I have found the following analogy helpful. Suppose a spaceship, the product of an advanced extraterrestrial civilization, lands on earth. Scientists begin to study the spaceship and find out it is a giant computer. First they study the hardware—the pieces of the computer—and try to determine how it is put together. After a while they see there is a logic in its design, and they invent theories of the computer—pictures of what kind of software programs can actually be processed by the computer—and these theories reflect the deep design invariances or laws of the computer.

Imagine next that the universe is a giant computer and what we see is the "hardware." The design of this computer is being discovered by physicists—these are their theories which tell us what programs can be run on the computer and can be checked by experiments. The physical laws are the invariances—the unchanging elements—in every possible program.

The computer-universe analogy can be carried too far. Once while visiting a laboratory near Moscow, I described the computer-universe analogy to a Soviet colleague and jestingly suggested that maybe what we perceive as the "hardware" of the universe is really "software" from another viewpoint. After briefly thinking this over my colleague confidently replied, "No software without hardware!" I agreed—form without content is meaningless.

I have been describing some of the characteristics of physical laws such as their universality, completeness, and relation to mathematics. However, looking for the physical laws is a human activity, and subjective, psychological elements enter. The conduct of scientific inquiry is not evident in the journals or lectures but is found in the discussions among a rather small group of researchers about the topics of current interest. Such discussion among scientists and the lonely periods of individual reflection and hard work are the creative moments that produce the insights crucial to science.

A predominant feature in the conduct of scientific research is intellectual aggression, which is manifested in the desire to know and to exercise one's intelligence in solving the enigma of the universe. No great science was discovered in the spirit of humility. A healthy sense of ego and intellectual intolerance is crucial to the conduct of inquiry. A colleague was complaining to me about the arrogant character of another theoretical physicist who had recently done some fine work. I replied, "No, you have it all backward. He has the marks of a messiah; he is a brilliant, self-confident, aggressive person who will appropriate others' ideas and think they are his own. He has a one-to-one ratio of ambition to ability." That such men can serve the search for truth while realizing their ambition comes as no surprise to scientists. It is not because humility is a liability that we should beware of it. It is because it is often fraudulent in a creative scientist, masking aggression. Once someone pointed out a very humble young physicist to Einstein and Einstein responded to the effect, "How can he be humble? He hasn't done anything yet!"

Physicists take their work seriously. If they did not, probably no one would, since it is so far beyond immediate human experience. But with that seriousness and commitment goes a remarkable sense of play. It is as if the universe were a joke or puzzle to play with. Without laughter and the joy of creativity the research enterprise would become unbearable. Humor opens the mind. It relieves the tension of concentration and exposes the vulnerability of a merely intellectual comprehension. Physicists love to make jokes about their work and its implications. It occurs to them that "the eternal Maker of enigmas" might also be a trickster.

I once heard a story that physicists when they die go to a heavenly academy where their purpose is to lay down the laws of nature. But there is a rule they must obey: Any new law they make cannot contradict ones already discovered and verified by their colleagues back on earth. The legend says that Pauli, one of the sharpest critics of physics, is there now setting intellectual traps and doing physics tricks to foul our best efforts!

To be a theoretical physicist one must be trained in mathematics and have good physical intuition. One cannot underestimate the role of intuition and imagination in the sciences. Students who do well on examinations do not necessarily become the most creative scientists. On an exam one is given a specific problem to solve, but in the real world of theoretical research the problem is to find the problem. Then it can be formulated precisely so that you can unleash your mathematical techniques upon it. Asking the right questions takes imagination.

Theoretical physicists swim in a sea of ideas. Which idea does one work on? As Einstein questioned, "If the researcher went about his work without any preconceived opinion, how should he be able at all to select out those facts from the immense abundance of the most complex experience, and just those which are simple enough to permit lawful connections to be evident?" That "preconceived opinion" is a crucial part of scientific inquiry—it is the partiality that guides the imagination to the relevant facts. As Richard Feynman, one of the inventors of quantum electrodynamics, remarked to me in his best New York accent, "To do physics ya gotta have taste." And taste, the instinct to work on the right problems, cannot be taught.

The imagination required to solve major problems in physics has an element of "craziness" in it, a well-grounded outrageousness or weirdness. Special relativity and quantum theory have that craziness. Once Pauli came to Pupin Laboratory at Columbia University to give a lecture upon Heisenberg's new nonlinear theory of elementary particles. Niels Bohr was in the audience, and after the lecture he remarked that the new theory couldn't be right because it wasn't crazy enough. Bohr and Pauli were soon standing on opposite ends of a table with Bohr saying, "It's not crazy enough," and Pauli responding with, "It's crazy enough." It would have been hard for an outsider to realize what was at stake for those two great physicists and that it wasn't simply madness. Both Bohr and Pauli knew that the craziness of the quantum theory turns out to be right.

All profound human creations are beautiful, and physical theories are no exception. An ugly theory has a kind of conceptual clumsiness which it is impossible to hold in the mind for too long. That is the basis for the appeal to aesthetics in the construction of physical theory. When physicists really understand the internal logic of the cosmos it will be beautiful—our attraction to the beautiful, what is coherent and simple, is at the heart of the human capability of rationally comprehending the material world.

While the final version of a profound physical theory is beautiful, it is a mistake to be motivated merely by the desire to construct a beautiful theory. The first time new ideas appear they are often bizarre and strange, and if the ideas are correct, beauty is seen later. Once when someone commented to Einstein that his general theory of relativity was very elegant, Einstein responded, quoting Ludwig Boltzmann, a physicist of a previous generation, "Elegance is for tailors!"

Beauty is in the eye of the beholder. The visual aesthetic of geometry appeals to some of us; to others beauty is the abstract world of symbols. In modern quantum physics the aesthetic sense is conceptual in contrast to an earlier time when physicists' ability to visualize the natural world played an important role. Instead of pictures we have symmetries described by mathematics. The quantum world of elementary particles is organized according to complex and beautiful symmetry principles.

The physicist's mind seeks symmetry. But having found it, it is quick to recognize the flaw in perfect symmetry. Rarely in nature are symmetries perfect. They are broken, often in a symmetrical way. It is this flaw in symmetry, like the requisite mistake in a Persian carpet, that attracts the mind and gives us new clues about the dynamics of the world. From the view of modern physics the entire world can be seen as the manifestation of a broken symmetry. If the symmetries of nature were actually perfect we would not exist.

From time to time in the sciences a true genius will emerge. I am not referring here to genius in technical power—that can be remarkable but sometimes quite superficial. A genius is someone who, like the ancient prophets, has a pipeline to the Godhead. It is a kind of madness, but it is correct.

The mathematician Mark Kac distinguishes two kinds of geniuses, those he calls ordinary geniuses and those he calls extraordinary or devious geniuses. An ordinary genius is someone just like you and me except that the genius's ability to concentrate, remember, and create is much greater than ours. Their creative reasoning can be communicated. Extraordinary geniuses are quite different. It is not at all clear how they think. They seem to work by a set of rules of their own invention and yet arrive at remarkable insights. They cannot tell you how they got there; their reasoning seems devious. The ordinary genius may have many students. But the devious genius rarely has any, since he cannot communicate his methods of solution.

Most scientists are not geniuses or not even near geniuses—but that need not inhibit their creativity or usefulness. The rules for creativity in science have never been written down and cannot really be learned from a book. Instead the conduct of inquiry is handed down from generation to generation of scientists, in a kind of charismatic chain—a teaching by example, not by the book. Being implicit, this tacit knowledge is easily altered by successive generations—an important but invisible aspect of scientific research.

Laying down the law in the physical sciences is a frustrating activity, an activity that promotes a sense of rational piety, a recognition that one is up against a major problem. I have always felt that Albrecht Dürer in his engraving *Melancholia* captured the spirit of rational inquiry. The engraving depicts a contemplative angel surrounded by the instruments of science, a magic square on a wall. It is an image of a consciousness whose isolation matches that of the stars.

2. The Cosmic Code

My friend, all theory is gray
and the golden tree of life is green.
—GOETHE, *Faust*

WHAT IS THE universe? Is it a great 3-D movie in which we are all unwilling actors? Is it a cosmic joke, a giant computer, a work of art by a Supreme Being, or simply an experiment? The problem in trying to understand the universe is that we have nothing to compare it with.

I don't know what the universe is or whether it has a purpose, but like most physicists I have to find some way to think about it. Einstein thought it a mistake to project our human needs onto the universe because, he felt, it is indifferent to those needs. Steven Weinberg agreed: ". . . the more we know about the universe the more it is evident that it is pointless and meaningless." Like Gertrude Stein's rose, the universe is what it is what it is. But what "is" it? The question will not go away.

I think the universe is a message written in code, a cosmic code, and the scientist's job is to decipher that code. This idea, that the universe is a message, is very old. It goes back to Greece, but its modern version was stated by the English empiricist Francis Bacon, who wrote that there are two revelations. The first is given to us in scripture and tradition, and it guided our thinking for centuries. The second revelation is given by the universe, and that book we are just beginning to read. The sentences within this book are the physical laws—those postulated and confirmed invariances of our experience. If there are those who claim a conversion experience through reading scripture, I would point out that the book of nature also has its converts. They may be less evangelical than religious converts, but they share a deep conviction that an order of the universe exists and can be known.

Many scientists have written about their first experience of contact with the cosmic code—the idea of an order beyond immediate experience. This experience often comes in the first years of adolescence when the emotional and cognitive life of an individual is integrated. Einstein said his conversion at that age from a religious to a scientific view of the universe changed his life. Newton, who held unorthodox religious views his whole life, also had a vision of the cosmic code—for him the universe was a great puzzle to be solved. I. I. Rabi, an atomic physicist, told me he first became interested in science when he took some books on the planetary motions out of a library. It was a source of wonder to him that the mind could know such immense things that were not invented by it. I myself remember as a teenager reading Einstein's biography, George Gamow's *One, Two, Three . . . Infinity,* and Selig Hecht's *Exploring the Atom* and making up my mind to become a physicist. I thought there was nothing more fulfilling I could aspire to than devoting my mind and energy to solving the puzzle of the cosmos. For me, physics, which explores the beginnings and ends of space, time, and matter, met those aspirations.

If we accept the idea that the universe is a book read by scientists, then we ought to examine how reading this book influences civilization. Scientists have unleashed a new force into our social, political, and economic development—perhaps the major force. By learning about the structure of the universe, scientists and engineers invent new technological devices which radically alter the world we live in. What distinguishes this new knowledge is that its source lies outside of human institutions—it comes from the material universe itself. By contrast, literature, art, the law, politics, and even the methods of science have been invented by us. But we did not invent the universe, the chemistry of our bodies, atoms, or electromagnetic waves—discoveries which profoundly influence our lives and history. Could it be that the cosmic code, revealed in the architecture of the universe, is actually the program for historical change?

Arnold Toynbee said that each civilization was a response to a challenge. The Romans had the challenge of maintaining dominion over a vast empire; their response was to invent the modern state. Likewise, the Egyptians met the challenge of their Nile environment by the construction of an elaborate irrigation system and a political structure to regulate it. The major challenge to our civilization is to master the discovered contents of the cosmic code. The forces science has discovered in the universe can annihilate us. They can also provide the basis of a new and more fulfilling human existence. What our response to this challenge will be no one knows, but we have clearly come to those sentences in the cosmic code that could bring our existence to an end or, alternatively, be the birth of humanity into the universe.

I complained once to an Indian friend that the poverty, ignorance, and hopelessness of the subcontinent were a consequence of Indian religious and philosophical beliefs (or was it the other way around?). My friend replied that some Indian intellectuals thought that the great wars of the West, wars which have taken millions of lives, are a consequence of Western philosophy, science, and technology. The challenge to our civilization which has come from our knowledge of the cosmic energies that fuel the stars, the movement of light and electrons through matter, the intricate molecular order which is the biological basis of life, must be met by the creation of a moral and political order which will accommodate these forces or we shall be destroyed. It will try our deepest resources of reason and compassion.

Our recent understanding also provides rich, complex, often confusing opportunities. We may feel that we exercise our freedom in the choices we make, but our options themselves are circumscribed by limits which have been made all the more clear by modern science. The condition of the universe, of the world, and of human life is viewed by many people as a product of science —rather than being seen as a discovery of science. It is a perception that results in a sense of alienation from the technological world.

In 1965 I was walking through the Boston Common with

friends and met an elderly woman with bright and lively eyes. She was wearing a handmade dress. A poet, she belonged to a small community which rejected the use of machines. (They wrote with quills.) The woman told me that her small group continued to believe in the human spirit but saw the human spirit as corrupted by modern life and by technology. She explained that a demonic spirit had come upon this earth about three hundred years ago, a spirit inimical to humanity, which it set out to destroy. The malevolence began when the best minds among the philosophers, scientists, and social and political leaders were captured. Soon the monsters of science, technology, and industrialism were loose upon the land. I thought of William Blake, another poet, lamenting Newton's blindness. The conquest was all but complete, she said; only a few held firm against the final fall.

The woman asked me what I did, and when I said I was a physicist I was greeted by a look of horror. I was one of "them," the enemy. I felt a chasm open between us. A year later the Counterculture was in full swing in America; a new revolt against science was on.

Some years later I spoke to a mentally disturbed young man. Very agitatedly, he described to me how alien beings from outer space had invaded the earth. They were formed of mental substance, lived in human minds, and controlled human beings through the creations of science and technology. Eventually this alien being would have an autonomous existence in the form of giant computers and would no longer require humans—and that would mark its triumph and the end of humanity. Soon he was hospitalized because he was unable to shake off this terrible vision.

The old poet and the young man are correct in their perception that science and technology came from "outside" the realm of human experience. They were sensitive to this perception in a way that most of us suppress. What is outside of us is the universe as a material revelation, the message that I call the cosmic code and that is now programming human social and economic development. What may be perceived as threatening in this alien contact is that scientists, in reading the cosmic code, have entered into the invisible structures of the universe. We live in the wake of a physics revolution comparable to the Copernican demolition of the anthropocentric world—a revolution which began with the invention of the theory of relativity and quantum mechanics in the first decades of this century and which has left most educated people behind. By the nature of the phenomena it studies, science has become increasingly abstract. The cosmic code has become invisible. The unseen is influencing the seen.

The irreversible transformation of the pattern of human existence by science is a profoundly disturbing experience that most people do not see because it is too close to them. Most of us live in huge cities with populations in the millions which simply could not have existed a few centuries ago because of the problems of supplying food and controlling disease. We accept technology as part of the structure of our lives because our survival depends on it. Experts and scientists assure us that technology is going to be all right because it is supported by the rule of reason. But others like the poet see reason as the tool of the devil, an instrument for the destruction of life and simple faith. They see the scientist as a destroyer of the free human spirit, while the scientist sees the poet's allies as blind to the material requirements of human survival. What divides us is the difference between those who give priority to intuitions and feelings and those who give priority to knowledge and reason—different resources of human life. Both impulses live inside each of us; but a fruitful coexistence sometimes breaks down, and the result is an incomplete person.

In the thirteenth century, scholasticism struggled to reconcile faith with reason. It failed, but out of its failure grew a new civilization—the modern world in which the dialectic between faith and reason continues to engage us. The dialectic is not to be resolved; it should be perceived as an opposition which transforms life. Our capacity for fulfillment can come only through faith and feelings. But our capacity for survival must come from reason and knowledge.

Is modern science hostile to our humanity? Max Born, one of

the developers of the quantum theory, expressed concern about the permanence of the scientific enterprise of the last three hundred years. Contemporary science, he felt, has no fixed and solid place in the constellation of human life as do politics, religion, or commerce. He wondered if humankind might ultimately abandon science. If that should happen, it would sever our still-fragile connection with the cosmic code—an error that might cost us our existence. I believe that future historians will see contemporary civilization as response to the discovery of the worlds of molecules, atoms, and the endless reaches of space and time. The challenge is to bring these invisible realms to consciousness and to make human the enormous powers we find there.

Science is another name for knowledge, and we have not yet discovered the boundary of knowledge, although we are discovering many other boundaries. But knowledge is not enough. It must be tempered with justice, a sense of the moral life, and our capacity for love and community. Science brings us to a renewed appreciation of the human condition—the limitations of our existence in the universe. Through the expansion of scientific awareness we learn again and again not only of further advances of our material possibilities but of their intrinsic limitations.

Genesis tells us about our first parents who were created in a garden paradise and made its stewards by the Lord. There were two trees, the tree of knowledge and the tree of life, and the Lord forbade them to eat of the fruit of the tree of knowledge. The first parents tasted knowledge and hence knew good and evil. They could now become, like the Lord, potentially infinite in knowledge. The Lord cast them from the garden before they could eat from the tree of life and become infinite in life as well. Humanity lives before the vision of infinite knowledge but from a state of finite being.

Science is not the enemy of humanity but one of the deepest expressions of the human desire to realize that vision of infinite knowledge. Science shows us that the visible world is neither matter nor spirit; the visible world is the invisible organization of energy. I do not know what the future sentences of the cosmic code will be. But it seems certain that the recent human contact with the invisible world of quanta and the vastness of the cosmos will shape the destiny of our species or whatever we may become.

I used to climb mountains in snow and ice, hanging onto the sides of great rocks. I was describing one of my adventures to an older friend once, and when I had finished he asked me, "Why do you want to kill yourself?" I protested. I told him that the rewards I wanted were of sight, of pleasure, of the thrill of pitting my body and my skills against nature. My friend replied, "When you are as old as I am you will see that you are trying to kill yourself."

I often dream about falling. Such dreams are commonplace to the ambitious or those who climb mountains. Lately I dreamed I was clutching at the face of a rock but it would not hold. Gravel gave way. I grasped for a shrub, but it pulled loose, and in cold terror I fell into the abyss. Suddenly I realized that my fall was relative; there was no bottom and no end. A feeling of pleasure overcame me. I realized that what I embody, the principle of life, cannot be destroyed. It is written into the cosmic code, the order of the universe. As I continued to fall in the dark void, embraced by the vault of the heavens, I sang to the beauty of the stars and made my peace with the darkness.

For the essence and the end
Of his labor is beauty, for goodness and evil are two things
and yet variant, but the quality of life as of death and of light
As of darkness is one, one beauty, the rhythm of that Wheel,
and who can behold it is happy and will praise it to the
people.*

* Robinson Jeffers, *Point Pinos and Point Lobos.*

Bibliography

The reader interested in pursuing some of the material discussed in this book should consult the references below. This bibliography contains but a small portion of the literature on modern physics—it is hardly complete but serves as an introduction. Almost all these books and articles are written by physicists, a fact reflecting my own bias in the selection. Some books, while they may use mathematics, all contain material accessible and of interest to the general reader.

Amaldi, Ugo. "Particle Accelerators and Scientific Culture." CERN reprint 79-06, Geneva 1979.

Asimov, Isaac. *Science, Numbers and I.* Garden City: New York, 1968.

Bernstein, Jeremy. *Einstein.* New York: Viking Press, 1973.

Bohr, Niels. *Atomic Physics and Human Knowledge.* New York: John Wiley and Sons, 1958.

Born, Max. *The Born-Einstein Letters.* New York: Walker and Co., 1971.

Childs, H. *An American Genius: The Life of Ernest Orlando Lawrence.* New York: Dutton, 1968.

Clark, Ronald W. *Einstein, The Life and Times.* New York: World Publishing, 1971.

Cline, Barbara L. *The Questioners: Physicists and the Quantum Theory.* New York: Thomas Y. Crowell, 1973.

Davies, P. C. W. *The Forces of Nature.* New York: Cambridge University Press, 1979.

———. *The Physics of Time Asymmetry.* Berkeley, California: University of California Press, 1974.

Dirac, P. A. M. "The Evolution of the Physicist's Picture of Nature." *Scientific American, 208,* May 1963.

Drell, S. D. "When is a Particle." *American Journal of Physics, 46,* June 1978, p. 597.

Einstein, Albert. *Ideas and Opinions.* Translated by Sonja Bargmann. New York: Crown Publishers, 1954.

———. *The Evolution of Physics.* With Leopold Infeld. New York: Simon and Schuster, 1938.

Feinberg, Gerald. *What is the World Made Of? Atoms, Leptons, Quarks and Other Tantalizing Particles.* New York: Doubleday, 1977.

Feynman, R. P.; Leighton, Robert B.; and Sands, Matthew. *The Feynman Lectures on Physics.* Vols. I, II, III, New York: Addison-Wesley Publishing Co., 1963.

———. "Structure of the Proton." *Science, 183* (1975).

Frank, Philipp. *Einstein: His Life and Times.* New York: Alfred A. Knopf, 1947.

Gamow, George. *One, Two, Three . . . Infinity.* New York: The Viking Press, 1947.

Gell-Mann, M., and Rosenbaum, E. P. "Elementary Particles." *Scientific American, 197,* July 1957.

Glashow, Sheldon. "Quarks with Color and Flavor." *Scientific American, 233,* No. 4 (1974), p. 38.

Heisenberg, Werner. *Physics and Beyond.* New York: Harper and Row, 1971.

Hoffmann, Banesh. *Albert Einstein, Creator and Rebel.* With collaboration of Helen Dukas, New York: The Viking Press, 1972.

———. *Albert Einstein: The Human Side.* With Helen Dukas. Princeton: Princeton University Press, 1979.

Holton, Gerald. "Constructing a Theory: Einstein's Model." *The American Scholar,* Vol. 48, No. 3 (1979).

———. *Thematic Origins of Scientific Thought: Kepler to Einstein.* Cambridge: Harvard University Press, 1973.

———. *The Scientific Imagination: Case Studies.* New York: Cambridge University Press, 1978.

Klein, Martin J. *Paul Ehrenfest.* Amsterdam: North-Holland; New York: American-Elsevier, 1970.

McMillan, E. "Early Accelerators and Their Builders." *IEEE Trans. Nuclear Science, 20,* June 1973.

Miller, Arthur I. *Albert Einstein's Special Theory of Relativity; Emergence (1905) and Early Interpretation (1905–1911).* Reading, Massachusetts: Addison-Wesley Publishing Company, 1981.

Pais, Abraham. "Einstein and the Quantum Theory." *Reviews of Modern Physics,* Vol. 51, No. 4 (1979).

Particles and Fields: Readings from *Scientific American.* With an introduction by William J. Kaufmann III. San Francisco: W. H. Freeman and Co., 1980.

Polkinghorne, J. C. *The Particle Play: An Account of the Ultimate Constituents of Matter.* San Francisco: W. H. Freeman and Co., 1979.

Rosen, Joe. *Symmetry Discovered.* New York: Cambridge University Press, 1975.

Schwitters, Roy F. "Fundamental Particles with Charm." *Scientific American, 237,* No. 4 (1977).

Schilpp, Paul A., ed. *Albert Einstein: Philosopher–Scientist.* Tudor Publishing Co., 1949.

Segre, Emilio. *From X-Rays to Quarks, Modern Physicists and Their Discoveries.* San Francisco: W. H. Freeman and Co., 1980.

Trefil, James S. *From Atoms to Quarks: An Introduction to the Strange World of Particle Physics.* New York: Charles Scribner's Sons, 1980.

Teller, Edward. *The Pursuit of Simplicity.* Malibu, California: Pepperdine University Press, 1980.

Weinberg, Steven. *The First Three Minutes.* New York: Bantam Books, 1979.

———."Unified Theories of Elementary Particle Interaction." *Scientific American, 231,* No. 1 (1974), p. 50.

———. "The Search for Unity: Notes for a History of Quantum Field Theory." *Daedalus, 106* (1977).

———. "The Forces of Nature." *American Scientist, 65* (1977).

Weisskopf, Victor W. *Knowledge and Wonder.* Cambridge, Massachusetts: MIT Press, 1979.

———. "Three Steps in the Structure of Matter." *Physics Today, 23,* August 1969.

———. "The Development of the Concept of an Elementary Particle." *Proceedings of the Symposium on the Foundations of Modern Physics,* Loma-Koli (Finland), 1977 (ed. V. Karimäki) B1, No. 14 (pub. University of Joensuu, 1977).

Wilson, R. "From the Compton Effect to Quarks and Asymptotic Freedom." *American Journal of Physics, 45* (1977).

Woolf, Harry, ed. *Some Strangeness in the Proportion.* Reading, Massachusetts: Addison-Wesley Publishing Company, 1981.

Index